Vetiveria
The Genus *Vetiveria*

Edited by

Massimo Maffei

Department of Plant Biology,
University of Turin,
Italy

CRC Press
Taylor & Francis Group
Boca Raton London New York

CRC Press is an imprint of the
Taylor & Francis Group, an **informa** business

CRC Press
Taylor & Francis Group
6000 Broken Sound Parkway NW, Suite 300
Boca Raton, FL 33487-2742

First issued in paperback 2019

© 2002 by Taylor & Francis Group, LLC
CRC Press is an imprint of Taylor & Francis Group, an Informa business

Typeset in 11/12pt Garamond 3 by Graphicraft Limited., Hong Kong

No claim to original U.S. Government works

ISBN-13: 978-0-415-27586-6 (hbk)
ISBN-13: 978-0-367-39638-1 (pbk)

British Library cataloguing in Publication Data
A catalogue record for this book is available from the British Library

Library of Congress Cataloging in Publication Data
A catalogue record has been requested

Visit the Taylor & Francis Web site at
http://www.taylorandfrancis.com

and the CRC Press Web site at
http://www.crcpress.com

Contents

List of Contributing Authors

Anand Akhila – Central Institute of Medicinal and Aromatic Plants. Lucknow, India 226 015. Telephone: 091 522 342676; Fax: 091 522 342666; email: akhiladr@lw1.vsnl.net.in

Cinzia M. Bertea – Department of Plant Biology, University of Torino, Viale P.A. Mattioli 25, I–10125 Torino, Italy. Telephone: +39 011 670 7447; Fax +39 011 670 7459; email: bertea@bioveg.unito.it

Wanda Camusso – Department of Plant Biology, University of Torino, Viale P.A. Mattioli 25, I–10125 Torino, Italy. Telephone: +39 011 670 7447; Fax +39 011 670 7459; email: camusso@bioveg.unito.it

Nwainmbi Simon Chia – Belo Rural Development Project (BERUDEP), PMB, P.O. Box 5, Belo-Boyo, N.W. Province. Cameroon.

Ruth E. Leupin – Institute of Biotechnology ETH Hönggerberg, CH–8093 Zürich, Switzerland. Telephone: +411 633 3286; Fax: +411 633 1051; email: leupin@biotech.biol.ethz.ch

Massimo Maffei – Department of Plant Biology, University of Torino, Viale P.A. Mattioli 25, I–10125 Torino, Italy. Telephone: +39 011 670 7447; Fax +39 011 670 7459; email: maffeima@bioveg.unito.it

Marco Mucciarelli – Department of Veterinary Morphophysiology, University of Torino, Viale P.A. Mattioli 25, I–10125 Torino, Italy. Telephone: +39 011 658 387; Fax: +39 011 670 7459; email: mucciama@bioveg.unito.it

Michael W.L. Pease – Co-ordinator, European and Mediterranean Vetiver Network. Quinta das Espargosas, Odiaxere, 8600 Lagos, Algarve, Portugal. Telephone and Fax: 351–82–798466; email: mikepease@mail.telepac.pt

Mumkum Rani – Central Institute of Medicinal and Aromatic Plants. Lucknow, India 226 015. Telephone: 091 522 342676; Fax: 091 522 342666.

Paul Truong – Principal Soil Conservationist, Resource Sciences Centre, Queensland Department of Natural Resources. Brisbane, Australia. Telephone: (07) 3896 9304; Fax: (07) 3896 9591; email: Truongp@dnr.qld.gov.au

Noel Vietmeyer – Office of International Affairs. National Academy of Sciences. Washington, DC 20418, USA. email: Noelvi@aol.com

Claudio Zarotti – Viale Teodorico, 2 20149 Milan, ITALY. Telephone: 02 – 324879; Fax: 02 – 325922; email: velasrl@tin.it

Introduction

Few plants have the capability of being both economically and ecologically important. *Vetiveria* possesses both qualities ensuring that the plants are one of the most versatile crops of the third millennium. The economic importance of the genus *Vetiveria* depends on the ability of the species *V. zizanioides* (L.) Nash to produce odorous roots, which can be used for the extraction of an essential oil of great economic importance. The ecological importance is due to the ability of the plant to act as a natural barrier against erosion and soil pollution.

The essential oil that is produced in secretory cells located inside the mature roots has been used since ancient times as a perfume as well as a natural remedy against human and animal diseases. It consists of a complex mixture of sesquiterpene hydrocarbons and alcohols, which are mostly used as a basic material for perfumery. The oil's aroma has a woody, earthy character; it is pleasant and persistent. The chemical composition of the oil makes vetiver oil an irreplaceable source when a woody, earthy note is required for a perfume. In this book, several chapters describe the characteristics of vetiver essential oil such as the site of oil production (Chapter 2) and oil biogenesis and chemical composition (Chapter 4). Pharmacological properties of *V. zizanioides* are described in a technical note in Chapter 5, and Chapter 3 deals with essential oil extraction including description of the distillation process and subsequent storage of the oil.

The economic importance of *Vetiveria* has prompted a series of biotechnological studies directed toward assessment of protocols for callus induction and plant regeneration as well as DNA analysis of the genome. These aspects are described in Chapter 7, where the authors present an overview of the most recent achievements in the field of vetiver biotechnology, as well as giving their own contributions to the cell, tissue and organ culture of the most important species of the genus, viz. *V. zizanioides*.

The ecological importance of the genus *Vetiveria* is described by some of the most qualified experts in this field. The ability of *V. zizanioides* to be a natural barrier against erosion cannot be separated from its characteristic of being an aromatic plant. During the last twenty years the importance of the plant as a simple and low cost method of soil conservation has grown, as well as its importance in producing a valuable essential oil. This growth of interest prompted the creation of The Vetiver Network, an international organization devoted to the spreading of the knowledge of the application of *V. zizanioides* for all ecological applications in what is called Vetiver Grass Technology. This topic, of sound worldwide importance, is described in Chapter 6, along with detailed description of other applications such as infrastructure and environmental protection.

Aspects of both ecology and oil production are considered in term of market trends, industrial needs, economy and environmental importance in Chapter 8, through a careful analysis of costs and benefits related to the vetiver global market. The last chapter analyses the future of vetiver, by giving a forecast of the needs and potential applications of the plants.

Vetiver oil production is still a source of income and a precious and irreplaceable source for basic notes to be used in perfumery. Vetiver grass technology is a natural force against erosion and for the fight against environmental pollution. It is really rare to find in a single plant such a wide range of applications and *Vetiveria* represent an almost unique example of a total-employment plant. In addition the plant is a high photosynthetic efficiency species, displaying C4 photosynthesis and is described in terms of anatomy, biochemistry and physiology in Chapter 2.

The aim of this book is to give the reader a full picture of the great potential of the genus *Vetiveria*, through a wide view on both economic (essential oil production and utilization) and ecological (vetiver grass technology) aspects. The contribution given by the authors, while giving an exhaustive and detailed state of the art on vetiver brings the connoisseurs of the properties of vetiver oil to the new world of vetiver grass technology. At the same time, it gives the users of vetiver as a "green line against erosion" the basic knowledge to appreciate the qualities of the plant as an essential oil producer.

1 Introduction to the Genus *Vetiveria*

Massimo Maffei

Department of Plant Biology, University of Turin, Viale P.A. Mattioli, 25 I–10125 Turin, ITALY

Historical Background

Vetiver grass, in particular the species *Vetiveria zizanioides* (L.) Nash, has been known to be a useful plant for thousands of years. It is mentioned in ancient Sanskrit writings and is also part of Hindu mythology. Rural people have used it for centuries for the oil from its roots, for the roots themselves, and for the leaves. Its center of origin appears to be in southern India and it has spread around the world through its byproduct value as a producer of an aromatic oil for the perfume industry. In the latter part of the last century and also in this century the sugar industry particularly in the West Indies, the off shore eastern African islands such as Mauritius and Reunion, and Fiji, has used the grass for its conservation properties (Grimshaw, 1998).

Vetiver grass has grown in the tropics over many centuries (NRC, 1993) and it has been mentioned among inscriptions on Kananj king copper plates since 1103 A.D. It has been cultivated longest for the scented oil produced by its roots as well as for the ability of the plant to retain soil and prevent erosion. For over two hundred years there has been an irresolved controversy over both the naming of the genus and the species of this grass. Hence, at the beginning of the twentieth century, the *Andropogon* grasses were frequently confused with each other and even after a great deal of library work and search for original specimens, as documentary evidence, in the older herbaria, Stapf stated that there could still be some confusion in the taxonomy of these grasses. Even today many doubts remain on the systematical identification of the species. Another source of doubt is raised by the fact that the cultivars of vetiver found in other parts of the world have been named individually by the different people; for example, the species called *V. nigritana* in Nigeria could in fact be *V. zizanioides*, and until we have some better means of distinguishing the species from each other the true identity remains doubtful (Greenfield, 1988).

Historically, vetiver grass was known by the peoples of Northern India by the popular names "Khas Khas" or "Vetiver." Other Sanskrit names which have been interpreted in the same sense are "Virana", "Lamajjaka" (or "Lamaja") and "Bala". The actual term used is 'turushka-danda', which Babu Rajendrala'la Mitra interprets as meaning 'aromatic reed' (turushka = aromatic substance, danda = stick), and hence also 'Khas Khas'. The latter term, now so commonly used, is supposed to be of Persian origin, but this appears doubtful (Greenfield, 1988). It has long been known that the roots, but not the leaves, are fragrant and are sold in the bazaars to prepare lotions, infusions and decoctions for medical purposes. In Colombo, Ceylon (Sri Lanka), the grass has been known since the seventeenth century as "Saewaendara", a name surviving to the present day. In 1700, it was known in Madras under the Tamil name of "Vettyveer" (=Vetiver), the vernacular name by which the grass is best known in Europe. A list of other common names is given later in this chapter.

Ground roots of vetiver have been used since ancient times to prepare odorous pads, while in woven form they provided perfumed strings, fans or curtains protecting from summer heat and producing a pleasant smell when watered or pushed by the wind. In infusion, roots have been used to provide a refreshing drink against fevers or stomach diseases. Used for topical applications, vetiver preparations are known to relieve pains caused by skin burns and warm sensations. The essential oil was used against cholera because of its emetic properties. Chapter 5 describes some of the main pharmacological properties of vetiver oil together with ethnopharmacological data. Nevertheless it is because of its odorous essential oil, used particularly in perfumery, that the plant is famous (Peyron, 1989).

Recently, many efforts have been made to use vetiver as an agent for soil erosion control and moisture retention. Vetiver was introduced into World Bank-assisted watershed projects in India for soil conservation on hillslopes. From the beginning, visual material demonstrating the effective use of hedges in Fiji helped immensely to convince Indian farmers. Despite resistance from the Indian Extension Service and research institutions and skepticism, vetiver demonstrations and trials grew and the technology met with success at the grassroots level, specifically through non governmental organizations (NGOs), technicians and farmers. The initial excitement and success with vetiver in India spread and now vetiver can be found incorporated into both on- and off-farm conservation programs in a growing number of countries including Australia, Bolivia, Brazil, China, Costa Rica, Ecuador, El Salvador, Guatemala, Honduras, India, Indonesia, Madagascar, Malawi, Malaysia, Mexico, Nepal, Nicaragua, Nigeria, Philippines, Sri Lanka, South Africa, Thailand, Zambia and Zimbabwe. Vetiver trials have begun in more than 25 countries as a result of World Bank extension efforts. Indeed, the excitement surrounding vetiver has grown so much that the potential for its use, both for agricultural and nonagricultural purposes, is exceptional (Slinger, 1997).

Description, Taxonomy, Distribution and Cultivation

Vetiveria Lem. Lisank is an important aromatic plant genus, belonging to the family Poaceae (Gramineae) (subfamily Andropogoneae) and comprising ten species. The genus is related to the genera *Sorghum* subgen *parasorghum* and *Chrysopogon* (Clayton and Renvoize, 1986). The species *V. zizanioides* and *V. lawsonii* are from the Indian subcontinent and *V. nigritana* is reported from Africa. *V. zizanioides* (L.) Nash, commonly known as vetiver grass or Khus grass, is widely distributed in India, Burma, Ceylon, and spread from Southwest Asia to tropical Africa (Bor, 1960). It is a coarse perennial grass, densely tufted, awnless, wiry and glabrous which occurs in both wild and cultivated forms in many parts of the tropics and subtropics.

The plant is a tufted perennial occurring in large clumps arising from a much branched "rootstock"; the culms are erect 0.5–3 m high (see Chapters 3 and 6 for illustrations). The leaves are basal and cauline with relatively stiff, elongate blades up to 80 cm long and 8 mm wide (Sreenath *et al*., 1994). The leaves are usually conduplicate basally, splitting along midrib apically and pubescent basally and sometimes purple.

Leaf margins are revolute, the uppermost bearing vitreous spines; sheaths are glabrous; the ligulesn narrow projections from the apices of leaf-sheaths of grasses, are fringe of hair 0.3–1 mm long. Panicles, loose irregular types of compound inflorescences common to grasses, are 15–30 cm long and comprise numerous racemose, spike-like branches and are usually purple with the rachis or stem disarticulating at base of sessile spikelet. The spikelets are dorsally compressed and paired, one being sessile and perfect, the other pedicellate and staminate or neuter. Some cultivated forms seldom flower. The sessile spikeleta are about 5 mm long, somewhat flattened laterally, bearing short sharp spines and are hermaphrodite with 3 stamens and 2 plumose stigmas each. The glumes or chafflike bracts of the sessile spikelets are also 5 mm long and are described as acuminate, coriaceous, nerveless, vitreous, papillose and spinose; the lemmas or bracts with flowers in their axes are about 3.5 mm long, acuminate, inrolled, tinged with purple and apices and margins are scarious (thin, dry and membraneous); the paleas or caryopsis is usually not seen. The pedicellate spikelets are slightly smaller than the sessile spikelets. The chromosome number is 2n = 20.

Other reputedly valid species and varieties belonging to the *Vetiveria* genus are: *V. arguta* (Steud.) C.E. Hubb., *V. elongata* (R. Br.) Stapf, *V. filipes* (Benth.) C.E. Hubb., *V. filipes* (Benth.) C.E. Hubb. var. *arundinacea* (Reeder) Jansen, *V. fulvibarbis* (Trin.) Stapf, *V. intermedia* S.T. Blake, *V. lawsoni* (Hook.f.) Blatter & McCann, *V. nigritana* (Benth.) Stapf, *V. pauciflora* S.T. Blake, *V. rigida* B.K. Simon and *V. zizanioides* (L.) Nash var. *tonkinensis* A. Camus.

Vetiveria zizanioides *technical specification*

Common names

INDIA – *Sanskrit*: Abhaya; Amrinala; Avadaha; Bala; Dahaharana; Gandhadhya; Haripriya; Indragupta; Ishtakapatha; Jalamoda; Jalashaya; Jalavasa; Katayana; Laghubhaya; Lamajjaka; Nalada; Ranapriya; Rambhu; Reshira; Samagandhika; Sevya; Shishira; Shitamulaka; Sugandhimula; Ushira; Vira; Virabhadra; Virana; Virataru; Vitanamulaka. *Hindi*: Bala; Balah; Bena. *Ganrar*: Khas; Onei; Panni. *Urdu*: Khas. *Bengali*: Khas-Khas. *Gujarati*: Valo. *Marathi*: Vala; Khas-Khas. *Mundari*: Birnijono, Sirum; Sirumjon. *Oudh*: Tin. *Punjab*: Panni. *Sadani*: Birni. *Santali*: Sirom. *Telugu*: Avurugaddiveru; Kuruveeru; Lamajjakamuveru Vettiveeru; Vidavaliveru. *Tamil*: Ilamichamver; Vettiver; Vilhalver; Viranam. *Kannada*: Vettiveeru; Laamanche; Kaadu; Karidappasajje hallu. *Mysore*: Ramaccham; Ramachehamver Vettiveru. SRI LANKA – *Sinhalese*: Saivandera; Savandramul. BURMA: Miyamoe. IRAN – *Persian*: Bikhiwala; Khas. CHINA – Xiang-Geng-Sao. MALAYSIA: Nara wastu; Nara setu; Naga setu; Akar wangi (fragrant root); Rumput wangi (fragrant grass); Kusu-Kusu. INDONESIA: Aga wangi; Larasetu; Larawestu; Raraweatu; Sundanese; Janur; Narawastu; Usar. PHILIPPINES: Ilib (Pamp.); Mora (Bik, Bis); Moras (Tag., Bis., Bi.); Moro (Tag), Narawasta (Sul); Raiz de moras (Sp); Rimodas (P. Bis); Rimora (Sbl); Rimoras (Bik); Tres-moras (Bis) Vetiver (Eng.); Amoora (C. Bis); Amoras (Ilk); Anias de moras (Pamp); Giron (P. Bis); Muda (Cebu – Central Visayas). LAOS and THAILAND: Faeg. SAHEL – *Bambara*: babin, ngongon, ngoko ba. *Songhai*: diri. *Fulani*: kieli, dimi, pallol. *Sarakolle*: kamare. *Mossi*: roudoum. *Gurma*: kulkadere. ETHIOPIA: Yesero mekelakeya. SENEGAL – Wolof: sep, tiep. *Falor*: toul. *Tuk*: semban. GHANA: N. Terr. Dag; kulikarili. N. NIGERIA – *Hausa*: jema. *Fulani*: so'dornde; so'mayo; chor'dor'de; ngongonari.;

zemako. SIERRA LEONE – *Mende*: pindi. *Susu*: barewali. *Temne*: an-wunga ro-gban. ARABIA: Izkhir. PUERTO RICO: Pachuli. FRANCE: Chiendent del Indes, Vetiver. U.K.: Cuscus, Khiskhus, Koosa (data from Greenfield, 1988).

Synonymy

Vetiveria zizanioides may be indexed in *Herbaria* under the following synonyms: *Andropogon zizanioides* Linn.; *Andropogon squarrosus* Hack; *Andropogon muricatus* Retz.; *Andropogon nardus* Blanco; *Andropogon nigritanus* Stapf.; *Andropogon festucoides* Presl.; *Andropogon echinulatum* Koenig; *Anatherum zizanioides* Linn.; *Anatherum muricatum* Beauv.; *Agrostis verticillata* Lam and *Phalaris zizanioides* Linn.

Habit, Soil and Climate

Although the plant grows in all kinds of soils, a rich well-drained sandy loam is considered best for harvesting the roots. The grass grows luxuriantly in areas with an annual rainfall of 1,000 to 2,000 mm at temperatures ranging from 21 °C to 44.5 °C although it can also grow at much higher and lower temperatures. *V. zizanioides* has been successfully established at 42° N. Lat., North of Rome, where it has survived snow for 18 days and −11 °C (including eight months when the minimum temperatures periodically dropped below freezing) at an altitude of 650 m with winter rainfalls of 1,100+ mm. It has been successfully established at an altitude of 2,300 m in the Himalayas (Pauri. UP. India), where it not only withstood extreme cold, but survived heavy grazing by goats, deer and other livestock on poor, eroded mountain soils. Wherever *V. zizanioides* has been planted it has grown. If grown for its essential oil, it requires its natural habitat of humid to sub-humid tropical conditions and alluvial or recent Andosols (volcanic ash soils that release the roots easily). By taking this plant from its natural habitat, where as a hydrophyte it put all its energy into seed production, and planting it in conditions ranging from the semi-arid tropics to the temperate zone, it functions like a xerophyte putting all its energy into a deep root system necessary for survival (Greenfield, 1988).

Habit: Perennial grass up to 2 m high with a strong dense and mainly vertical root system often measuring more than 3 m. It is by nature a hydrophyte, but often thrives under xerophytic conditions.

Climate: Temperature – Mean 18–25 °C; Mean coldest month 5 °C; Absolute minimum −15 °C. When the ground freezes the grass usually dies. Growth normally starts again above 12 °C. Hot summer temperatures (25 °C+) are required for rapid growth.

Rainfall: as low as 300 mm, but above 700 mm preferable; the plants will survive total drought, but normally require a wet season of at least three months. Ideal conditions include a well spread monthly rainfall.

Humidity: although the plants grow better under humid conditions, they can also do well under low humidity conditions.

Sunshine: the plants are difficult to establish under shade; however, when shade is removed, growth recovery is rapid.

Soil: the plants grow best in deep sandy loam soils. However, plants will grow on most soil types ranging from black cracking vertisols through to red alfisols. They will grow on rubble, comprising both acid (pH 3) and alkaline (pH 11) soils, and are tolerant to high levels of toxic minerals e.g. aluminium and manganese (550 ppm).

These grasses will survive complete submergence in water for up to three months and will grow on both shallow and deep soils.

Altitude: plants will grow at altitudes up to and over 2,000 m, growth being constrained by low temperatures at higher altitudes.

Distribution

According to Greenfield (1988) the distribution of *V. zizaniodes* is as follows:

India, Southeast Asia (Thailand, Malaysia, Philippines, Indonesia), Pakistan, Polynesia (Samoa, New Caledonia, Fiji, Tonga), Nepal, Tropical Africa, as well as South Africa, Burma, Sri Lanka, Guyana, New Guinea, French Guyana, Argentina, West Indies (Haiti, Cuba, Jamaica, Puerto Rico, Antigua, St. Vincent, Martinique, Barbados, Trinidad), Colombia (Santa Maria), Brazil (Rio de Janeiro, Para, Bahia), Paraguay (Central Paraguay), China (Fujian and Jiangxi provinces; Hainan Island), Zimbabwe (introduced early 1960s from Mauritius); Kenya (introduced to Kenya in November 1987); Somalia (introduced to NWRADEP Hargeisa, April 1987); Nigeria (introduced to Kano, October 1987); United Kingdom (introduced in November 1988); Italy (introduced to Migliorino Pisano at the end of twentieth century).

Selection and cultivation

There is a growing awareness of the need to understand the taxonomy of vetiver, and how vetiver is related to species and cultivars within the genus. DNA testing offers a unique opportunity to test the relationships. Kresovich *et al.* (1993, 1994) showed the value of this when DNA testing indicated that the non-flowering Huffman Boucard accessions were closely related to the vetivers found in Guatemala and Jamaica, and were very different from the seeding cultivars introduced from north India by the ARS.

In order to establish a vetiver grass nursery, Greenfield (1988), in his report on vetiver grass to the World Bank, offered a series of recommendations which are worth quoting: "To find a source of vetiver grass, first check with the local herbarium (located in the university, or botanical gardens, or agricultural department) to determine if it has any specimens of *V. zizanioides*. If they have, withdraw the specimen sheet, and in the bottom lefthand corner there should be a small map showing where this particular specimen was collected. This will show you what the plant looks like; give you the locality where it was found, and if the collection was done correctly, it will tell you the local name of the plant . . . Assuming you have a source of planting material, dig out the clumps of vetiver, cutting the roots off about 20 cm below the surface. Cut off the leaves about 30 cm above the roots, and break the clump into planting pieces, or 'slips,' of about five tillers per slip, taking care to discard dead or seeded tillers. Single tillers, (shoots of the plants springing from the bottom of the original stalk), will suffice if you are desperately short, but it is better to plant a small clump. The nursery is best located in an irrigated field, which will encourage the plants to grow very rapidly. Prepare the nursery bed as you would for any field crop: plow it, cultivate it, and get rid of the weeds. The seed bed need not be smooth, as the vetiver seedlings (slips) are extremely hardy. Irrigate the plot thoroughly, then transplant the slips as you would transplant rice, except the vetiver slips are spaced

40 cm apart. It does not matter if it rains after planting and inundates the slips. The plant will not be affected. This wide spacing gives each plant ample room to 'tiller,' or produce more planting material. There is no set planting distance and with experience you may develop a planting distance suited to local conditions. The planting material should be brought to the nursery at least six months before the planting season so that it is available at the beginning of the wet season. Once all the slips have been planted in the nursery, fertilizer can be applied if necessary. Plants grow faster and produce more tillers in less time if they are fertilized. Generally we use phosphate fertilizers in combination with some form of nitrogen; for instance, sulphate of ammonia super phosphate, or urea super phosphate, or diammonium phosphate, whatever is available and cheap in the nitrogen-phosphate range. In India, farmers use manure on the nursery beds. The more optimum amounts of fertilizer applied according to the needs of the soil, the more planting material will be produced in six months. In the first two months, when the plants are getting established, weed the beds to keep the weeds under control. Once the plants have started to grow vigorously, keep them trimmed to about 50 cm, and use the cut leaves to mulch between the rows and keep the weeds down. Trimming encourages 'tillering' and produces more planting material in a shorter period. If the plants are allowed to flower, tillering is reduced. After six months there should be between 80 and 100 tillers per plant, which can be used as planting material. Thoroughly soak the plants to make it easier to lift them out of the ground. Quite often, it takes a two-man team using a strong fork, or pick, or even a bar (crowbar) to remove the tillers from the ground. One man levers the plant out of the soil, the other pulls the top of the plant toward him. Once sufficient roots are exposed, they can be cut 20 cm below the surface, and pulled out by the other team member. Now the clump can be broken up into planting pieces for transport to the field. When harvesting, leave three or four tillers in the ground from each clump to renew the planting material. Fertilize and irrigate the remaining plants, using all the trash from the harvest to mulch the beds. Succeeding harvests may be possible in four to five months. Transporting the material to the field is no problem. Trim the slips as stated above – 30 cm of leaves/20 cm of roots, both trimmed with a machete – put them in grain bags or throw them in the back of a truck. The plants can stand a lot of rough handling, and can be left unattended for 10 days. It is always better if you can plant them the same day, but if they have to be transported over great distances or stored, the losses will be negligible. Planting should be done at the very beginning of the wet season."

Pests and Diseases

Vetiver is propagated vegetatively and there is no evidence of the plant being invasive under upland rainfall conditions (NRC, 1993). One of the main objectives of the National Research Council's review (NRC, 1993) of Vetiver was to verify whether vetiver might be a threat as a potential weed. The review found that in the majority of instances vetiver was not invasive, but it strongly recommended that only the non-seeding accessions be used. Evidence suggests that accessions from southern India are less prone to seeding than those from northern India. There are reports that accessions introduced from India to ARS stations in Mississippi were very fertile and germinated strongly. More research is required into the flowering habits of vetiver in relation to cultivar, climate, rainfall, and day length (Grimshaw, 1989).

Vetiver is extremely resistant to insect pests and diseases (Yoon, 1991; Zisong, 1991). There is evidence from India that when dead vetiver plant material is affected by termites there may be an allelopathic reaction that prevents regrowth of vetiver from the center of the plant, and under severe drought conditions, new young shoots on the periphery of the plant are grazed out and the plant is killed. Alternatively, and most probably, the termite cast is too tough for the new young shoots to penetrate. Management by burning may eradicate this problem (Grimshaw, 1989). Reports from Brazil (York, 1993) suggest that vetiver is resistant to the race 1 (root knot nematodes) *Meloidogyne javanica* and *M. incognita*, both serious root nematodes in tobacco. Two *Vetiver* varieties were screened for resistance against five root-knot nematode populations. The populations are representative of the main genetic groups of *Meloidogyne* in Australia and consist of four species (viz. *M. arenaria*, *M. incognita* [populations B1 and B2], *M. javanica* and *M. hapla*). They were identified using DNA technology. The two *Vetiver* varieties used were Monto, a sterile selection, and a non-sterile type from Western Australia. Both *Vetiver* varieties were highly resistant to all five root-knot nematode populations. Reproduction was approximately 1,000-fold less than on the susceptible tomato. Vetiver also compared favorably with other grasses that have been found to be resistant to root-knot nematodes in similar tests (West *et al.*, 1999). Since the grass was resistant to all major species of *Meloidogyne*, vetiver is unlikely to exacerbate problems caused by root-knot nematodes when used as a cover, companion or hedgerow crop. In China there have been reports that vetiver has been affected by the rice stem borer (Wang and Zisong, 1991), and although this has not effected the growth of the vetiver, the latter might act as a host plant. However in Fujian (south eastern China), where vetiver has been grown in close association with rice for many years, this does not seem to be a problem. In most cases pests and diseases in vetiver can be best controlled through burning, and, as will be noted later in this chapter, burning may have an important place in the general management of vetiver hedges. An investigation of insects on vetiver hedges was carried out in the Guangxi Province of China (Shangwen, 1999). In 211 days of investigation, 1,225 common species of insects were found in the Nanning suburb of Guangxi Province; of these there were 79 species on vetiver hedges. These insects came from 53 families of 13 orders. There were 14 species from Hymenoptera, 11 species from Orthoptera, 10 from Diptera, 9 from Homoptera, 9 species from Lepidoptera, 9 from Coleoptera, 5 from Odonata, 4 from Hemiptera, 3 from Mantodea, 2 from Blattaria, 1 from Isoptera, 1 from Dermaptera, and 1 from Collembola (Shangwen, 1999). Of 79 species of insects on vetiver hedges, insects eating vetiver leaf in test tube tests were locust, snoutmoth, tussockmoth and leaf beetle. Owing to a tiny population density and the limited amount eaten these insects did not cause any negative effect on vetiver growth. On the contrary, beneficial predators (total 30 species) were attracted to the vetiver hedges; species such as mantids, dragonflies, ladybugs etc. are all important predator insects of garden, agriculture, and forestry pests. Therefore, when vetiver is introduced to a new environment it would be appropriate to practice integrated pest management (IPM) (Shangwen, 1999). Evidence indicates that, overall, vetiver is resistant to pest and diseases, and is not seen as a serious host plant (Grimshaw, 1989).

Feed Value

There has been very little research carried out on the management and feed value of vetiver as a fodder, and vetiver grass is normally ignored by grazing livestock. In Malaysia sheep will not eat vetiver in the field when there is an abundance of other more palatable species, but cut tops when fed to penned sheep were readily consumed. In China and Malaysia vetiver has been successfully fed to grass carp. In eastern Indonesia, under very dry conditions, vetiver was eaten by cows and horses. Under good management young vetiver leaves have a nutritive value similar to napier grass with Crude Protein levels of about 7.0%. Under good conditions high volumes of green leaf are available. There is little doubt that with some improved management vetiver would make an adequate dry season fodder, particularly if combined with a high protein forage. Farmers at Gundalpet, India, have been using vetiver for centuries as a field boundary, and during the peak growing season it is cut once every three weeks for fodder. Reports of its use as a fodder come from many other countries including China, Guatemala, Honduras, Niger and Mali. Some accessions are known to be more palatable e.g. the so-called "farmer" cultivar from Karnataka, which had been selected by farmers over decades as a softer and more palatable cultivar (Grimshaw, 1989). See Chapters 3 and 8 for pictures and more data on the feeding value of vetiver grass.

In areas where there are more palatable species of forage grass or where livestock are absent, users who require an inert grass that can be developed with minimum management should look to vetiver. There are excellent examples of this application demonstrated in Costa Rica (Grimshaw, 1993) for the protection of mango orchards on steep slopes. Whether vetiver will be used as a fodder will be determined by the management objectives of the user. One thing required is the identification and screening of accessions that are more palatable and manageable as a dual purpose conservation and forage plant.

Essential Oil Production

Vetiver oil is one of the perfumers' most basic traditional materials. It possesses fixative properties that help to render long lasting the effects of the composition in which it is used. The oil's aroma is basically of a heavy, woody, earthy character, pleasant and extremely persistent (Sreenath et al., 1994). It is difficult to reproduce with synthetic aromatic chemical formulations because of the complexity of the molecular structures. No identical synthetic substances can be found in commerce. Chemical research has produced a set of generally woody smelling molecules, more or less recalling the typical vetiver smell e.g. vetiver cedar smell Vertofix (widely used) and others with a woody-cedar note; various vetiver-like notes (quinolein, nootkatone, floro-pal, fhubofix, etc.) and other woody subtances such as b-isolongifolene (Peyron, 1989).

Elite germlines of *Vetiveria zizanioides* (L.) Nash have long been cultivated for their fragrant roots, which contain the essential Oil of Vetiver. This oil is clearly distinguished chemically and in commerce from Khus oil, which comes from natural (fertile) populations of *V. zizanioides* in the Ganges Plains of North India (CSIR, 1976). The Oil of Vetiver has long been produced pantropically through Vetiver cuttings. Within the past decade, vetiver occurrence has increased enormously through widespread plantings (over 100 countries) to form hedges for stabilizing soil and controlling waterflow.

Vetiver oil is used both in fine perfumery and in a whole range of soaps, skin lotions, deodorants and other cosmetic applications. Occasionally it is the dominant contributor to a fragrance; more often it is used as a solid foundation upon which other fragrance notes are superimposed (Sreenath *et al.*, 1994). Commercially vetiver essential oil has two main applications, firstly as the pure oil and secondly as the acetate (totally esterified) derivative. The term vetiverols refers to the alcohol-rich portion of sesquiterpenoids present in the oil. The major economically important oils are those originating from Reunion, Haiti, Angola, Guatemala, China, Java and Brazil. Vetiver oil is steam distilled from the roots which possess a most agreeable odor (see Chapter 3 for more details on oil distillation). Lemberg and co-workers were the first to study the chemical composition of some vetiver oils of different geographical origin (Lemberg and Hale, 1978). Their investigation raised some of the main problems encountered in the chemical analysis of this complex oil. They found that no single one step method of analysis would suffice for the total delineation of the components. Gas chromatography is still the major method of analysis, providing extensive data, but only after preliminary efficient fractionation of the crude oil. A typical analysis of the oil indicates the presence of three major fractions: the lower boiling or sesquiterpene hydrocarbon fraction, the intermediate fraction containing the bulk of oxygenated derivatives and the third fraction represented by the so-called "Khusimol" fraction, which contains the main component alcohol in the vetiver oil (Lemberg and Hale, 1978). The sesquiterpene fraction dominates the oil composition. Based on comparative studies we can distinguish a group of dextrorotatory oils derived from *V. zizanioides* Stapf. (from South India, Java, Haiti, Brazil, Reunion) which are particularly rich in eudesmane, epieudesmane, nootkatane, spirovetivane and zizaane derivatives. A group of levorotatory oils derived from *V. zizanioides* Linn. from two areas north of India are particularly rich in enantiomorph cadinanes and norsesquiterpene derivatives, with a khusilane structure and a selinane structure. Oils from Angola are obtained from *V. nigritana* Benth (Peyron, 1989).

Another important method of analysis for vetiver essential oil is organoleptic analysis, which allows subjective evaluation of the oil value and quality. Other methods include multivariate statistical analyses such as Principal Component Analysis (PCA) and Discriminant Analysis (DA). PCA is the basic method for data analysis; it consists in describing data concerning n subjects to whom p characteristics are associated by means of two or three special axes know as principal components. This analysis has been used to allow graphic displays of subjects and characters, pointing out similarities between subjects, which can be then classified in homogeneous groups. DA allows testing the previously defined groups and creates tables of predict versus group data. Furthermore it determines a new character, a logical combination of old characters, so that the new one is as discriminating as possible. Vetiver oil biogenesis will be described in detail in Chapter 4.

Vetiver oil and its derivatives show many olfactory facets, thus representing a precious basic material as mentioned above. The oil smell is defined by a dozen olfactory describers, strictly corresponding to and providing suitable images. The favourable olfactory describers are woody (liquorice wood, safety matches wood, warm, rich, powerful, wide, etc.) and earthy (root, soil, moist underwood, potatoes sack, toasted peanuts, truffle, light-green head note, etc.), whereas unfavourable olfactory describers are burnt (too long toasted peanuts, guaiol, cooked, boiled, etc.), chemical product (carbide, oil, kerosene, cedar, caryophyllene, sawdust, guaiacol, etc.) and green (greenbeans, too strong potato peel, etc.) (Peyron, 1989).

The various olfactory facets of vetiver oil are used for the preparation of various male perfumes among which can be mentioned Aramis, Monsieur Rochas, Eau de sport Lacoste and Patou pour homme. Male perfumes which have vetiver in their names are Vétiver Carven, Eau de Vetiver le Gallion, Vétiver Guerlain, Vetyver Lanvin, Vetyver Roger & Gallet, Vetiver de Puig and Bois de Vetiver Bogart. Among feminine perfumes we find Amazone, Chanel 22, Or noir, Soir de Paris, L'heure bleu, Arpège, Madame Rochas, Vivre, Calandre, Azzaro, Crèpe de Chine, Bandit, Miss Dior, Anais Anais, First, Diorella, etc. (Peyron, 1989).

In classical toilette waters (e.g. Guerlain), the bergamot bitter freshness is followed by the earthy-woody constituent of vetiver oil. Carnation vibrates the woody note and celery accentuates the sulfurous one. In Chanel No. 5, jasmin and may rose warmth reach their utmost thanks to the vetiver velvet note.

The distillation of oil from vetiver roots is a long, delicate and expensive operation and a careful description of methods of distillation is given by Zarotti in Chapter 3 of this book. The differences in the quality of the oil, according to the producing country, depend on the various technologies used. Peyron (1989) in his review of techniques of vetiver extraction for perfumery, gives a clear picture of the different methodologies used for the extraction of vetiver oil from roots. In northern India clay ovens heated with fung can be found situated beside communal ponds. The upper part consists of a large copper boiler with a bamboo tube as the swan neck. Condensates are collected in copper bottles with tapered necks, situated in the cool pond, which acts as a refrigerator. Oil collection is performed by keeping the floating oil. In South India the technology is based on a series of modern stainless steel alembics. In Java dry distillation is performed by steam injection with a stainless steel diffuser, whereas in the Reunion Islands steam for distillation is often provided by waste oils fed into the boiler. In Congo (Kinshasa) machine rasped roots are successively kept soaking in water overnight, whereas in Brazil an efficient distillation apparatus can be found in Campinas with an original condenser from which bent and curved rib shaped stainless tubes depart downwards facilitating dense oil flow. The oil separator is formed in two parts, an inner part for light and intermediate fractions and an outer part for heavy fractions (Peyron, 1989).

The production of the essential oil has in recent times faced no difficulty in keeping pace with demand and the main sources of vetiver oil in the 1980s were Haiti, Indonesia (Java), Reunion, Brazil, China and, to a lesser extent, India (Robbins, 1982). The oil has also been produced in other countries: the Indian subcontinent, the Far East and South and Central America, but not in sustained quantities. A general estimate suggests that annual world production of vetiver oil was about 200–250 tonnes in the 1980s (Robbins, 1982).

In terms of qualitative composition, the best oil is produced on the Reunion Islands, but in terms of quantity of production the market is dominated by Haiti and Indonesia, the Haitian oil being distinctly better than the Indonesian (Sreenath et al., 1994). The annual average Haitian production of vetiver oil is around 100 tonnes (all exported) and the same average amount is produced in the island of Java. The essential oil produced on Reunion is widely known by the appellation "Bourbon", the annual "Bourbon" production was estimated as 20–30 tonnes of which three quarters are exported (Robbins, 1982); however since 1960 the production has declined sharply from 50 tonnes to a mere 6 tonnes in 1989. The principal producers now are Haiti and Java, while newcomers on the vetiver oil market are China and Brazil (Guadarrama, 1990).

In the 1980s the annual consumption of vetiver oil was dominated by USA (100 t), followed by France (50 t), Switzerland (30 t), U.K. (25 t), Japan (10 t) and other countries (30–40 t), with the price of the oil on the New York market in 1991 being as follows: Bourbon 136.70 US $/kg, Chinese 37.50 US $/kg and Javanese 35.30 US $/kg (Robbins, 1992; Sreenath et al., 1994). According to Sreenath et al. (1994), the reputation of vetiver oil has suffered primarily as a result of market distortion and the rather indifferent and variable quality of some oils (Indonesia). Bourbon oil has tended to lose favor primarily because of its high price. The impossibility of formulating an artificial vetiver essential oil at a realistic price tends to reduce the consumption of vetiver oil in new products, forcing oil utilizers to move towards alternative materials of a woody character such as cedarwood oil and its derivatives.

Several investigations have been performed on vetiver oil describing new techniques of oil analysis, oil extraction, oil component structure determination and oil biosynthesis during development as well as exploration of the potential of vetiver oil as an insect deterrant or as a pharmacologically bioactive mixture of compounds (Smadja, 1994).

Other Uses of Vetiveria

Vetiver against soil erosion

Beside essential oil production, vetiver has been successfully used to prevent soil erosion. It can be planted to protect the edges, or along the waterline, especially on curves of unlined canals. It protects both the upper and lower slopes of secondary and tertiary canals as well as aqueducts leading back from the main canal and around the foothills to the upper reaches of the command area. It can protect the bunds of rice paddies as well. Vetiver grass hedges around the perimeter of a dam will prevent silt from washing into the dam, and give the dam a much longer useful life. It can protect the inlet to the dam by filtering the silt out of the water before it enters the dam. In China, some farmers use it not only to stabilize the wall of the dam, but as a fence to keep ducks from getting out of the dam. The grass has many other uses to stabilize soil. To the engineer, grasses are invaluable for stabilizing sand dunes, road verges, and other raw soil surfaces. Orchards have shown remarkable response to being planted behind a vetiver hedge on very steep slopes (>100%). In Trinidad, mango trees planted behind the vetiver hedge have outgrown the trees planted away from it. By filtration of the run-off, the trees obtain the full benefit of run-off products. When the rains stop and the dry season sets in, the hedges can be cut to the ground and their leaves used as a mulch around the base of the trees. Vetiver mulch is very long lasting. The leaves also make an excellent thatch. In Fiji it is considered the best thatching material, and lasts for at least three years. Vetiver grass in India and Zimbabwe is used to keep rhizomatous weeds out of fields. The rhizomes of *Cynodon dactylon* cannot penetrate the deep curtain of vetiver roots and farmers say that once all the rhizomes are dug out, they have no more problems. In Zimbabwe, it has been used to protect tobacco fields from couch grass (Greenfield, 1988; Grimshaw, 1989, 1998). This paragraph outlines some of the most important uses of *V. zizanioides* that are not related to essential oil production. For a deeper insight into vetiver grass technology see Chapter 6. A careful cost analysis can be found in Chapter 8.

The magnitude and effect of soil erosion is astounding. On a global scale exact rates of soil erosion are unknown and difficult to measure; however, estimates point to a possible 10 to 20 billion tons of soil lost each year worldwide, representing the equivalent loss of between 5 million and 7 million hectares of arable land (NRC, 1993). Other than limited application for soil and moisture conservation, there are other problems such as vegetative barriers, that a technology must face once it is on the ground (Slinger, 1997). Vetiver's unique physical characteristics gives the grass distinct advantages that are beneficial for soil and moisture conservation. It can be utilized in the form of dense, vegetative contour barriers that can reduce the velocity of running water and spread the water out, increasing the plant available moisture. At the same time the hedge diminishes the movement of soil down the slope. The result can be the development of a soil terrace structure behind the hedge. Through conserving soil and moisture the vetiver hedge acts to maintain the fertility of the soil and thereby raise its productivity. Compared with elephant grass (*Pennisetum purpureum*), vetiver gives neighbouring crops little or no competition and the vertical structure of the roots allows associated crops to grow right up to a vetiver hedge, seemingly without interference and loss of yield (NRC, 1993) (see also Chapter 3). John Greenfield and Richard Grimshaw introduced vetiver into World Bank-assisted watershed projects in India. Greenfield had been involved in using vetiver in Fiji in the 1950s for soil conservation on hillslopes and Grimshaw was, at that time, chief of the Agricultural Division of the World Bank in India. From the beginning, visual material demonstrating the use of the hedges in Fiji helped immensely to convince Indian farmers, particularly as the Fiji farmers on the tape related their experiences in Hindi. Despite scepticism and resistance from the Indian Extension Service and research institutions, vetiver demonstrations and trials grew and the technology met with success at grassroots level, specifically through NGOs, technicians and farmers. The initial excitement and success with vetiver in India spread and now vetiver can be found incorporated into both on- and off-farm conservation programs in a growing number of countries including Australia, Bolivia, Brazil, China, Costa Rica, Ecuador, El Salvador, Guatemala, Honduras, India, Indonesia, Madagascar, Malawi, Malaysia, Mexico, Nepal, Nicaragua, Nigeria, Philippines, Sri Lanka, South Africa, Thailand, Zambia and Zimbabwe. The diffusion of vetiver technology to these other countries, particularly for soil and moisture conservation, can largely be attributed to the efforts of Richard Grimshaw. Within his work in some of these countries Grimshaw incorporated vetiver as he saw the need. Vetiver trials have begun in more than 25 countries as a result of World Bank extension efforts (Slinger, 1997).

Vetiver and soil conservation

For a plant to be useful in soil conservation it must have certain characteristics. For example, it must be capable of forming a dense, permanent hedge, resistant to the harmful effects of overgrazing and fire; it must be perennial and permanent, being capable of surviving as a dense hedge for centuries; its crown must be below the surface, to protect it from fire and overgrazing; it must be sterile, also producing no stolons or rhizomes so it will not become a weed and furthermore it should repel rodents, snakes, etc. These characteristics apply to *V. zizanioides* which is a remarkable plant and no other is known to share its hardiness or diversity. The sharp leaves and aromatic roots of vetiver keep it free of vermin and other pests. Vetiver is also both a xerophyte and a hydrophyte and, once established, is not affected by droughts or floods (Greenfield, 1988).

Sometimes vetiver is the only effective technology for soil and moisture conservation on very steep slopes. Experience in Honduras has indicated that on slopes above 30%, species other than vetiver were not sufficiently effective to allow for sustainability. On the other hand, using vetiver contour barriers cropping could be sustained on slopes of 60% (see also Chapter 3 and Chaper 6). Within a year and a half vetiver grass was successfully used to stabilize more than 200 km of El Salvadorian roadway. The recognition of the importance of vetiver technology in road construction and embankment building has made it a major subject of an international Bio-Engineering conference held in the Philippines in 1999.

There is a move to promote methods of soil and moisture conservation that not only do the job of stopping soil erosion but also provide the farmer with animal fodder, fuel, food and a range of other products. One of the greatest criticisms of vegetative hedges, especially vetiver, as methods of soil and moisture conservation is that they do not have the multipurpose capacity of tree hedges or other agroforestry systems. Today there is a broadening of vetiver usage in two ways: (1) the recognition of multiple functions of the grass have been in use for centuries and (2) using vetiver for purposes other than for soil erosion control. The multipurpose features of vetiver include its use as fodder for animals, as mattress stuffing, as animal bedding, as mulch, in medicinal remedies, as thatch for roofs and as a material for making saleable woven handicrafts such as baskets and hats (Slinger, 1997). Vetiver grass may not provide a crop for the farmer but it can improve the soil and raise the productivity of the associated crops. In the long run, the farmer may need to incorporate a number of these methods into his/her management strategy. Farmers should be encouraged to compare techniques and discover for themselves the best solutions to their soil conservation problems. Vetiver grass technology has potential across a wide range of soils and climates. It also has the advantages of being one of the least intrusive and demanding technologies for soil and moisture conservation. Importantly, vetiver is also compatible with other methods of soil erosion control. Vetiver is already being planted in several countries to reinforce and improve the stability of terraces, berms (narrow ledges) and bunds (embarkments). It can also be used to restore the soil to a point where it is able to support other soil conservation measures such as alley cropping. Vetiver technology is often the only appropriate technology on slopes greater than 30%. These advantages make vetiver a necessary and important part of any menu of methods for soil and moisture conservation. International leading institutions, through their funding of specific programs, have the ability to ensure that the menu of soil conservation alternatives offered to small farmers is diverse (Slinger, 1997). Chapter 6 describes the vetiver hedge technology.

Vetiver and soil restoration

Vetiver grass (*Vetiveria zizanioides L.*) is widely used worldwide for soil erosion and sediment control (Truong, 1993). Research in Queensland showed that vetiver is highly tolerant to drought and water logging, frost ($-11\ °C$) and heat ($>45\ °C$), extreme soil pH (3.3 to 9.5), sodicity (ESP = 33%), salinity (17.5 mS cm^2 for 50% yield), aluminum toxicity (>68 Al/CEC%), manganese toxicity (>578 ppm) (Truong and Claridge, 1996). Vetiver is also highly tolerant to a range of heavy metals such as arsenic, cadmium, copper, chromium and nickel (Truong and Claridge, 1996). Vetiver is suitable for the stabilization and rehabilitation of acid sulfate soils. When adequately supplied with essential nutrients, vetiver produces excellent growth under highly acid conditions (pH = 3.8) with an extremely high level of soil aluminum and vetiver growth occurs at a level much higher than 68% aluminum saturation. Young leaves emerge and remain green for three weeks after planting in a soil with 87% aluminum saturation. Vetiver is more tolerant to aluminum toxicity than some of the most tolerant crop and pasture species such as rice (>45%), corn (30%), wheat (30%), soybean (20%), lucerne (15%) and cotton (10%) (Fageria et al., 1988).

Heavy metals are generally found naturally only at very low concentrations. Elevated concentrations are commonly associated with pollution from human activities. Heavy metals can affect plant growth by interfering with enzyme activities or preventing the absorption of essential nutrients. Many plants are sensitive to heavy metals, but those which are tolerant are generally tolerant of most heavy metals.

Vetiver yield was significantly reduced when the soil arsenic level was at 250 ppm or higher. Although such results did not establish the exact toxic threshold level for vetiver, it is most likely between 100 and 250 ppm; this level is extremely high when compared with the threshold of 0.02–7.5 ppm of the heavy metals used in trials under hydroponic conditions (Bowen, 1979). As arsenic soil levels at 20 ppm and 100 ppm may require rehabilitation vetiver is a highly suitable species for reclaiming these sites (Truong and Claridge, 1996). Vetiver growth is not significantly affected until the soil chromium exceeds 200 ppm and the toxic threshold level is between 200 and 600 ppm. This level is extraordinarily high when compared with the threshold of between 0.5 and 10.0 ppm of hydroponic conditions (Bowen, 1979). Vetiver is therefore highly suitable for reclamation of chromium contaminated lands as sites with soil chromium levels at 50 ppm may require rehabilitation. The critical copper toxic level for vetiver is between 50 and 100 ppm, while vetiver is highly suitable for the rehabilitation of nickel contaminated lands containing 60 ppm of nickel or higher. According to Baker and Eldershaw (1993) the toxic threshold level of nickel in the soil is between 7 and 10 ppm for most plants.

In Australia vetiver has been successfully used to stabilize mining overburden and highly saline, sodic, magnesic and alkaline (pH 9.5) tailings of coal mines and highly acidic (pH 2.7) and high arsenic tailings of gold mines (Truong, 1999). The important implications of these findings are that when vetiver is used for the rehabilitation of sites contaminated with high levels of arsenic, cadmium, chromium and mercury, its shoots can be safely grazed by animals or harvested for mulch as very little of these heavy metals are translocated to the shoots. As for copper, lead, nickel, selenium and zinc, vetiver usage for the above purposes are limited to the thresholds set by the environmental agencies and the tolerance of the animal concerned. In addition, although vetiver is not a hyper-accumulator it can be used to remove some heavy metals from the contaminated sites and disposed of safely elsewhere, thus gradually reducing the contaminant levels. For example vetiver roots and shoots can accumulate more than 5 times the chromium and zinc levels in the soil (Truong, 1999).

Rehabilitation trials conducted on both tailings dumps and slimes dams at several sites, have shown that vetiver possesses the necessary attributes for self sustainable growth on kimberlite spoils. Vetiver grew vigorously on the alkaline kimberlite, containing run-off, arresting erosion and creating an ideal micro-habitat for the establishment of indigenous grass species. Rehabilitation using vetiver was particularly successful on kimberlite fines at Cullinan mine where slopes of 35° are being upheld. It is clear that vetiver is likely to play an increasingly important role in rehabilitation and, as a result of this, nurseries are being established at several mines (Knoll, 1997). Chapter 7 emphasizes the role of vetiver in soil rehabilitation as well as its importance in phytoremediation and environmental protection.

The Vetiver Network

In 1989, while still in the employment of the World Bank, Richard Grimshaw created the Vetiver Information Network, later converted in 1995 to a non-profit organization with a Board of Directors and independently audited accounts through which information pertaining to vetiver can be disseminated, collected and re-networked. Communications for the Network are made via a publication, the Vetiver Newsletter, over the internet and on its website page (http://www.vetiver.org). The initial audience, as well as the database, for this Vetiver Network came from World Bank counterparts, their partners and associated research stations throughout the world. The core network consisted of 500 to 800 institutions, many involved in natural resource management and rural development. In 1997 the network comprised over 4,000 contacts in about 100 countries, with unknown numbers of potential contacts through re-networking. The Vetiver Grass Network has also been able to address instances of management problems and educate its audience about proper treatment of material. Slide shows with scripts are sold on a non-profit basis. The Vetiver Network serves to put potential users in contact with others in their country and to answer technical questions.

Since the Vetiver Information Network was established the clientele of the newsletters has broadened from being mainly foreign government officials and consultants to reaching NGOs, researchers in educational and government institutions and occasionally farmers.

Grimshaw has also been instrumental in the establishment of seven national and regional Vetiver Networks. Using donations from the Royal Danish Government, The World Bank, The Amberstone Trust and most recently, individuals, institutions and corporations, as well as the receipt in 1996 of the prestigious $100,000 John Franz Sustainability Award donated by the Monsanto Company, the Vetiver Information Network has provided financial support to a number of NGOs who are working with vetiver, supporting vetiver workshops, supporting and providing awards for vetiver research and providing basic funding for the establishment of regional and national vetiver networks (Slinger, 1997).

The Network's strategy is to help develop and establish regional and national networks that can in the future lead the drive in the dissemination of the technology quickly and efficiently to potential users in their area of influence. In particular The Vetiver Network will transfer hard copy production and delivery of newsletters to up and running national and regional networks.

At present there are seven regional and national Vetiver Networks comprising networks in China, Europe, Latin American, the Philippines, Southern Africa, the Pacific Rim and West Africa. The Filipino Network has three subregional networks. The Pacific Rim network was set up and is financially supported by the Thai Royal Family. All of the networks were set up during 1996 and 1997. Most of the regional and national networks were developed through the passion and interest of individuals who had some experience with vetiver and who recognized the potential of the technology as an alternative for soil and moisture conservation (Slinger, 1997).

In the future the Vetiver Network hopes to rely on the national and regional networks to identify suitable non-profit agency recipients for funding. Another benefit of the regional and national networks that has not yet been realized is their potential for seeking their own external funding for the establishment of research trials, demonstration plots and to support local NGOs and educational institutions experimenting with vetiver. Presently, the work for these networks is done on a voluntary basis; time and financial constraints will not allow for the further pursuit of these activities.

The Vetiver Network has played a key role in marketing the technology, linking users and researchers together, and providing feedback on a worldwide basis. The Vetiver Network currently operated from Leesburg, Virginia, USA should continue because we need a center to assure that information is exchanged and that contacts are made. The Vetiver Newsletter should be continued both in hard and electronic (Internet) copy. If funding is weak then the Internet way would be the cheapest. If the Vetiver Network is able to continue to source funds, it would be possible to move some of those funds to enable local networks to initiate their own operations (Grimshaw, 1998).

DNA Studies on Vetiver

One of the desirable features of hedgerow essential oil-producing vetiver is that it is non-fertile and so it must be propagated from cuttings. However, the mere fact that it is always distributed by cuttings could lead to the widespread cultivation of a single clone. This could be extremely dangerous. An insect or disease adapted to a particular genotype could spread and decimate millions of erosion control terraces of vetiver. In order to investigate this concern, Robert Adams and Mark Dafforn assembled leaf materials from around the world and compared these accessions to known wild and related materials using Random Amplified Polymorphic DNAs (RAPDs) (Adams and Dafforn, 1997).

Genetic variability was initially investigated by Kresovich *et al.* (1993, 1994), who reported on vetiver variation in the United States. They found RAPD patterns were very stable within clones, that the non-fertile "Huffman" and "Boucard" cultivars were identical (>.99+), and that these were clearly distinct from the USDA PI 196257 seed introductions from northern India (Simla, Punjab). Interestingly, they found that three samples of this USDA accession (#s 1,2,3), though similar, were genetically distinct from one-another. They concluded that RAPDs would be useful for identifying truly distinct sources of genetic diversity.

Based on several accessions (n = 121) of vetiver (*V. zizanioides* (L.) Nash) and related taxa from its region of origin and around the world, Adams and Dafforn (1997) concluded: "it appears only one *V. zizanioides* genotype, 'Sunshine', accounts for almost all germplasm utilized outside South Asia. Curiously, no 'Sunshine' types were detected from within this region of vetiver's early distribution. Additional RAPD analyses revealed that at least seven other non-fertile accessions are distinct genotypes. This germplasm diversity holds promise for reducing the vulnerable genetic uniformity in what is now essentially a pantropical monoculture of an economically and environmentally important plant resource. Evaluation trials of these accessions are planned. DNA from air-dried leaves was often found to be degraded beyond use (n = 22). Material submitted for DNA analysis should be small (actively growing) leaves, harvested fresh and immediately placed into activated silica gel or other suitable drying agent." The challenge is to assure the genetic diversity of cultivated vetiver, which is proving of immense importance to agricultural stabilization and civil engineering (Adams and Dafforn, 1997). More data on vetiver biotechnology, genetic variability and DNA analysis are given in Chapter 7.

Acknowledgements

I would like to thank very much Dick Grimshaw who kindly gave me much of the reference material for this chapter and for the revision of the manuscript. The help of Dr Ellena in giving data on vetiver in perfumery is also acknowledged.

References

Adams, R.P. and Dafforn, M.R. (1997) DNA fingerprints (RAPDs) of the pantropical grass vetiver, *Vetiveria zizanioides* (L.) Nash (Gramineae), reveal a single clone, "Sunshine", is widely utilized for erosion control. *TVN Newsletters*, 18.

Baker, D.E. and Eldershaw, V.J. (1993) Interpreting soil analyses for agricultural land use in Queensland. *Project Report Series Q093014, QDPI*, Brisbane, Australia.

Bor, N.L. (1960) *Grasses of Burma, Ceylon, India and Pakistan*, Pergamon Press, Oxford.

Bowen, H.J.M. (1979) *Plants and the Chemical Elements*, Academic Press, London.

Clayton, W.D. and Renvoize, S.A. (1986) *Genera graminum, grasses of the world*, HMSO Books, London.

CSIR (Council on Scientific and Industrial Research) (1976) *Vetiveria*. In *The Wealth of India*, Vol. X. Publications & Information Director, CSIR, New Delhi, pp. 451–457.

Fageria, N.K., Baligan, V.C. and Wright, R.T. (1988) Al Toxicity in Crop Plants. *Journal of Plant Nutrition*, 11, 303–307.

Greenfield, J.C. (1988) Vetiver grass (*Vetiveria* spp.) The ideal plant for vegetative soil and moisture conservation. World Bank, Washington D.C., June 23. *TVN Newsletters*, 1.

Grimshaw, R.G. (1989) The Role of Vetiver Grass in Sustaining Agricultural Productivity. *TVN Newsletters*, 2.

Grimshaw, R.G. (1993) Soil and Moisture Conservation in Central America, Vetiver Grass Technology, Observations from Visits to Panama, Costa Rica, Nicaragua, El Salvador, Honduras, and Guatemala. July 4–16 1993. Asia Technical Department, The World Bank, Washington, D.C.

Grimshaw, R.G. (1998) Vetiver Grass – An International Perspective. http://host.fptpday.com/vetiverbeta/International.htm

Guadarrama, D. (1990) The decline of vetiver production in Reunion. *Marchés Trop. Méditerr.*, 46, 71–77.

Knoll, C. (1997) Rehabilitation with vetiver. *African Mining*, 2, 2.

Kresovich, S., Lamboy, W.F., Li, R., Ren, J., Szewc-McFadden, A.K. and Bliek, S.M. (1993) Application of Molecular Diagnostics for Discrimination of Accessions and Clones of Vetiver Grass. *TVN Newsletters*, 10.

Kresovich, S., Lamboy, W.F., Li, R., Ren, J., Szewc-McFadden, A.K. and Bliek, S.M. (1994) Application of molecular methods and statistical analyses for discrimination of accessions and clones of vetiver grass. *Crop Science* 34, 805–809.

Lemberg, S. and Hale, R.B. (1978) Vetiver oils of different geographical origins. *Perfumer and Flavorist*, 3, 23–27.

NRC (National Research Council) (1993) *Vetiver grass: A thin green line against erosion*. Board on Science and Technology for International Development. National Academy Press, Washington, D.C.

Peyron, L. (1989) Vetiver in perfumery. *Quintessenza*, **13**, 4–14.

Robbins, S.R.J. (1982) Selected markets for the essential oils of patchouli and vetiver. *Report of the Tropical Products Institute*, London, G 167, 56.

Shangwen, C. (1999) Insects on Vetiver Hedges. *TVN Newsletters*, **20**.

Slinger, V. (1997) Spreading the slips of vetiver grass technology: A lesson in technology diffusion from Latin America. *TVN Newsletters*, **18**.

Smadja, J. (1994) Mise au point bibliographique sur l'huile essentielle de vetyver. *EPPOS*, **14**, 15–36.

Sreenath, H.L., Jagadishchandra, K.S. and Bajaj, Y.P.S. (1994) *Vetiveria zizanioides* (L.) Nash (Vetiver grass): *in vitro* culture, regeneration, and the production of essential oils. In Y.P.S. Bajaj (ed.), *Medicinal and Aromatic Plants VI*, Springer-Verlag, Berlin, pp. 403–421.

Truong, P.N.V. (1993) Report on the International Vetiver Grass Field Workshop, Kuala Lumpur. *Australian Journal of Soil and Water Conservation, 4*, 23–26.

Truong, P.N.V. (1999) Vetiver grass technology for mine tailings rehabilitation, First *Asia-Pacific Conference on Ground and Water Bio-engineering, Manila April 1999*.

Truong, P.N.V. and Claridge, J. (1996) Effects of heavy metals toxicities on vetiver growth. *TVN Newsletters*, **15**.

TVN Newsletter – The Vetiver Network Newsletters published biannually by The Vetiver Network, 15 Wirt St. N.W., Leesburg, VA 20176, USA (ftp://www.vetiver.org/Newsletters/Nlcontents.htm).

West, L., Stirling, G. and Truong, P. (1999) Resistance of Vetiver grass to infection by root-knot nematodes (*Meloidogyne* Spp). *TVN Newsletters*, **20**.

Yoon, P.K. (1991) A look-see at Vetiver grass in Malaysia – First Progress Report. *TVN Newsletters*, **6**.

York, P.A. (1993) Is there a role for vetiver grass on tobacco farms? *Zimbabwe Tobacco Association Magazine*, **2**, 6.

Zisong, W. (1991) Excerpts From The Experiments and Popularization of Vetiver Grass, Nanping Prefecture, Fujian Province, China. *TVN Newsletters*, **6**.

2 Anatomy, Biochemistry and Physiology

Cinzia M. Bertea and Wanda Camusso

Department of Plant Biology, University of Turin, Viale P.A. Mattioli 25, I–10125 Turin, Italy.

Introduction

Vetiver is a perennial graminaceous plant (Poaceae = Gramineae), originary from India, growing wild, half-wild or cultivated in many tropical and subtropical areas. Its fragrant roots contain essential oils used in the perfumery and cosmetic industry (Peyron, 1989). The name derives from the Tamil "vetti" (khus-khus or cus-cus) and "ver" (root), alluding to aromatic roots. Various botanical names can be found in the literature: *Vetiveria zizanioides* (L.) Stapf.; *V. zizanioides* (L.) Nash.; *V. muricata* Griseb.; *Andropogon muricatus* Retz.; *A. squarrosus* Hack; *Anatherum zizanioides* (L.) Hitchachase; *V. nigritana* Benth.

Ten species are known and their distribution includes tropical Africa (*V. fulvibarbis, V. nigritana*), Asia (India, in particular) (*V. lawsoni, V. arguta, V. nemoralis*) and Australia (*V. elongata, V. filipes, V. intermedia, V. pauciflora, V. rigida*). However, although vetiver cultivation is possible in all regions with a rather warm and moist climate, it is intensively cultivated in many semitropical areas including Guatemala, Brazil, Haiti, Angola, Somalia, Congo, Reunion Island, India, China and Indonesia (Peyron, 1989). The cultivation of vetiver in these regions is mainly aimed towards oil extraction, since vetiver oil is widely used in perfumery as a basic element in perfume blends (Sethi and Gupta, 1960). The essential oil has a complex and, until now, a relatively unknown composition, including for the most part sesquiterpenes and sesquiterpene alcohols (Lemberg and Hale, 1978; Smadia *et al.*, 1986, 1988; Akhila and Thakur, 1989).

The plant is herbaceous, unbranched above, forming large clumps from stout rhizomes. It grows widely, with culms 0.5–3 m high bearing sheathing leaves. It blooms only in some tropical areas but not in temperate regions. The inflorescence (25–45 cm long and 10–15 cm wide) is paniculate, brown or reddish, with male and hermaphrodite flowers, typically in pair, it is formed by a number of both sessile and pedunculate spikelets.

The root system is wide, consisting of long, fibrous roots and rootlets going down more than 2 m deep. About 85% of the roots can be found in the first 30–35 cm under the soil surface, where it creates a fasciculated mass formed by 5 to 30 cm long roots and rootlets of 1–2 mm diameter (Peyron, 1989). The extensive root system grows straight down, without interfering with the growth of neighbouring crops and anchors the plant firmly to the ground, stabilizing soil even on steep slopes with large amounts of water run-off. These features make vetiver grass a formidable tool against soil erosion and for this reason it is employed by the US government in bioremediation of eroded lands and in soil conservation (National Research Council, 1993; Dalton, 1994; Tscherning *et al.*, 1995; Mucciarelli *et al.*, 1997).

The combination of these economical and ecological properties prompted several research programs aimed at the utilization of the genus *Vetiveria* for environmental application.

Despite the ecological and economical importance, little modern research has been done on vetiver from a physiological point of view. Most of the knowledge of the genus *Vetiveria* is related to essential oil composition and soil erosion prevention. At present, information on the photosynthetic mechanism is limited. However, this chapter describes the results of several experiments conducted on *Vetiveria zizanioides* Stapf. plants collected by Sacco (1960) in Somalia and cultivated for several years at the Botanical Gardens of the University of Turin. Anatomical, biochemical and physiological approaches have been used in order to determine the photosynthetic mechanisms and to evaluate the potential growth capabilities of this plant in humid-temperate climates.

Anatomy

General considerations

Higher plants can be divided into two groups, C_3 and C_4, based on the mechanism utilized for photosynthetic carbon assimilation and related to anatomical and ultrastructural features. Photosynthesis by C_3 plants involves only one photosynthetic cell type, and in these plants atmospheric CO_2 is fixed directly by the primary carbon fixation enzyme ribulose 1,5-bisphosphate carboxylase/oxygenase (Rubisco). In contrast, C_4 plants, such as the monocotyledonous maize, possess a Kranz-type anatomy consisting of two cell types, mesophyll and bundle sheath cells, that differ in their photosynthetic activities (Hatch, 1992; Maurino *et al.*, 1997).

The C_4 pathway is a complex adaptation of the C_3 pathway that overcomes the limitation of photorespiration and is found in a diverse collection of species, such as different tropical and subtropical grasses, many of which grow in hot climates. About half of the species of the Poaceae are included among the C_4 plants (Smith and Brown, 1973). The key feature of C_4 photosynthesis is the compartmentalization of activities into two specialized cell and chloroplast types. Rubisco and the C_3 photosynthetic carbon reduction cycle (PCR) are found in the inner ring of bundle sheath cells. These cells are separated from the mesophyll and from the air in the intercellular spaces by a lamella that is highly resistant to the diffusion of CO_2 (Hatch, 1988). Thus, by virtue of this two-stage CO_2 fixation pathway, the mesophyll-located C_4 cycle acts as a biochemical pump increasing the concentration of CO_2 in the bundle sheath an estimated 10-fold over atmospheric concentrations. The net result is that the oxygenase activity of Rubisco is effectively suppressed and the PCR cycle operates more efficiently.

C_4 plants have two chloroplast types, each found in a specialized cell type. Leaves of C_4 plants show extensive vascularization, with a ring of bundle sheath (BS) cells surrounding each vein and an outer ring of mesophyll (M) cells surrounding the bundle sheath. CO_2 fixation in these plants is a two-step process.

There are three variations on the basic C_4 pathway and the biochemical distinctions are correlated with the ultrastructure differences of Kranz cells (Gutierrez *et al.*, 1974;

Hatch *et al.*, 1975). The three C_4 variants can be distinguished ultrastructurally by using a combination of two characters of bundle sheath cell chloroplasts, the degree of granal stacking and chloroplast position (Gutierrez *et al.*, 1974).

For this reason comparative grass leaf anatomy in relation to photosynthesis along with biochemical studies has become the object of intensive investigation.

Leaf anatomy

The first anatomical studies of the genus *Vetiveria* were undertaken by Vickery (1935) on *V. elongata* C.E. Hubbard and by Prat (1936, 1937) on *V. zizanioides.* Further studies on the latter species were conducted by Metcalfe in 1960. The descriptions given by these authors are very similar and still valid. The genus *Vetiveria* possesses a typical grass leaf (45–100 cm long; panicoid-type) which consists of a more or less narrow blade (0.6–1.2 cm wide) and a sheath enclosing the stem. Leaves 21–30 veined and rather thick, are rigid and hard in the part nearest the stem and become more tender towards the tips, where they are usually curved. Both leaf surfaces are more or less flat. The abaxial surface is slightly undulating, with ribs over the large and medium-sized vascular bundles.

The abaxial epidermis of *V. zizanioides* contains a variety of cells. The ground mass comprises narrow, elongated cells with moderately thick, sinuous walls, both over and between the veins. Abundant paired short-cells are also present both over and between the veins. It is possible to observe silica bodies over the veins, most crystals being cross shaped with very short arms. Microhairs (39–54 μm in length), often balanoform, are visible on the abaxial epidermis, while prickle-hairs, angular and very large, are present at the leaf margins. Figure 2.1a shows an electron scanning micrograph of *V. zizanioides* leaf blade. It is possible to observe developed prickle-hairs directed towards the apex of the leaf, which make the leaves extremely sharp. However, the genotypes which are cultivated most often are those with less sharp leaves.

The abaxial epidermis also reveals several stomata (27–33 μm long) arranged in rows, with narrow guard cells associated with subsidiary cells, which are often dome-shaped or triangular. Figure 2.1b represents an electron scanning micrograph showing a *V. zizanioides* stoma. This plant possesses typical grass stomata, with guard cells narrow in the middle and enlarged at the end (Figure 2.1c).

Cross section of the blade

The mesophyll of grasses shows, as a rule, no distinct differentiation into palisade and spongy parenchyma, although sometimes the cell rows beneath both epidermal layers are more regularly arranged than in the rest of the mesophyll. In *V. zizanioides*, the mesophyll cells surround the vascular bundles in an orderly manner owing to the chlorenchyma being chiefly confined to a comparatively narrow strip subjacent to the abaxial epidermis, with extensions along the sides of the medium-sized and large vascular bundles. The greater part of the lamina is occupied by very large, intercellular cavities extending from the assimilatory tissue to just below the adaxial epidermis. In Figure 2.1d is reported a cross section of *V. zizanioides* young leaf, excised from the shoot apex, illustrating the development of air-cavities. Most part of the mesophyll is still present. Cross section of *Vetiveria* old leaf is reported in Figure 2.1e. The picture shows well developed air cavities, lined by large thin-walled cells without chloroplasts.

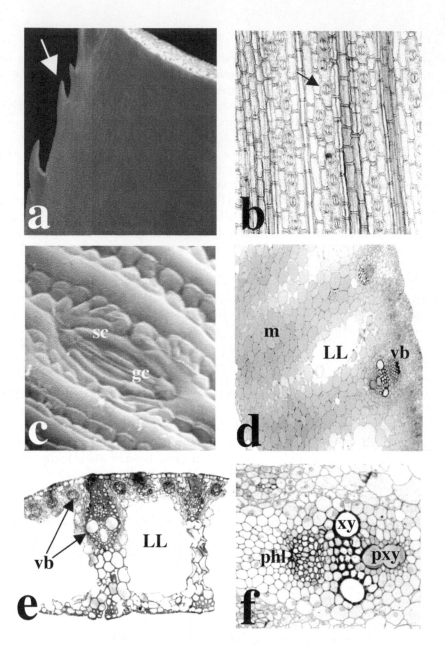

Figure 2.1 a Electron scanning micrograph of *V. zizanioides* leaf blade. Prickle-hairs are evident on the leaf edges (white arrow) (80x). *b* Abaxial epidermis in surface view showing stomata (black arrow) arranged in regular lines (from Watson and Dallwitz, 1999) (125x). *c* Electron scanning micrograph of a stoma seen from the surface. Guard cells (gc) are narrow in the middle and enlarged at the end, while subsidiary cells (sc) are dome-shaped (2000x). *d* Cross section of a young leaf developing lysigenous lacunae. Most part of the mesophyll is still present (134x). *e* Transverse section from an old leaf showing a single-layered epidermis on both sides of the blade and well developed lysigenous lacunae, lined by large thin-walled cells without chloroplasts. The vascular bundles (vb) of various sizes are delimited from the mesophyll by a single bundle sheath (134x). *f* Enlarged view of a large vascular bundle in which one bundle sheath is present. Details: phl, phloem; xy, xylem; pxy, protoxylem lacuna (336x).

A single-layered epidermis is present on both sides of the blade. The vascular bundles of various sizes are delimited from mesophyll by a single bundle sheath, more evident in the smaller ones.

V. zizanioides possesses bulliform cells which are confined to a single large group on the adaxial surface of the midrib. Colourless cells, besides those forming girders, and those in the adaxial part of the lamina towards the leaf margins, constitute the portion of the midrib ground tissue between the adaxial bulliform cells and the assimilatory tissue in the keel. During excessive loss of water, the bulliform cells, or the colourless cells (hinge cells, according to Esau, 1977), or both types in conjuction, become flaccid and enable the leaf to fold or to roll. In fact *V. zizanioides* leaves become folded on drying (Metcalfe, 1960).

Figure 2.1f shows a large vascular bundle at higher magnifications in which it is possible to identify a well developed protoxylem lacuna (pxy) along with the phloem (phl) and the bundle sheath extension.

Leaf ultrastructure

The structure of parenchymatic bundle sheath (BS) is particularly important in distinguishing C_3 and C_4 grasses. A commonly mentioned anatomical feature of C_4 plants is the orderly arrangement of mesophyll cells with reference to the bundle sheath, the two together forming concentric layers around the vascular bundle.

In contrast to the sheath cells, that have thick walls, the mesophyll cells have thin walls and intercellular spaces between them.

Bundle sheath chloroplasts contain starch grains in the stroma between the simple, agranal, internal membranes. In contrast, mesophyll chloroplasts possess grana.

In C_4 plants, the sheath has a high content of organelles, especially mitochondria and microbodies, and its chloroplasts are commonly larger than the mesophyll ones.

Both types have chloroplasts, ribosomes, plastoglobuli and the usual double membrane envelope.

A rather consistent character of chloroplasts in both sheath and mesophyll is the peripheral reticulum, a system of anastomosing tubules contiguous with the inner membrane of the chloroplast envelope (Esau, 1977).

An important characteristic of BS chloroplasts is also their position; they can be centrifugal or centripetal. Comparative biochemical studies on C_4 plants suggest that the variation in morphology and localization of the sheath chloroplasts are correlated with different enzymatic activities (Gutierrez *et al.*, 1974).

The anatomical differences between plants exhibiting a C_4 photosynthetic carbon assimilation pathway are disclosed by light and ultrastructural observation of the BS (Gutierrez *et al.*, 1974; Hatch *et al.*, 1975; Chapman and Hatch, 1983; Edwards and Walker, 1983; Hatch, 1987; Jenkins *et al.*, 1989). In the NADP-ME species chloroplasts are peripherally arranged and grana are deficient or absent in BS cells. A further distinctive feature is the presence or absence of the "mestome sheath", a layer of cells between the metaxylem vessels and adjacent BS cells which is present in NAD-ME and PCK species only (Hattersley and Watson, 1976). Moreover, the mitochondrial profile area in NADP-ME type species is lower than in NAD-ME ones, where the enzymes involved in the transformation of aspartate to CO_2 and pyruvate are present in the mitochondria (Hatch *et al.*, 1975). Another distinctive criterion is the presence of suberized lamellae, which can occur in the BS cell walls of several NADP-ME and

PCK species (Eastman *et al.*, 1988). Our results point to an evident NADP-ME bundle sheath Kranz cell anatomy for *V. zizanioides*. The absence of granal thylakoids and peroxisomes in BS Kranz cells indicates the absence of O_2 evolution and the presence of a photorespiratory pathway (Cramer *et al.*, 1991).

Light microscopical observations of *V. zizanioides* leaves revealed the presence of Kranz BS cells common to the majority of C_4 species (Laetsch, 1974). In *V. zizanioides* cross sections of minor veins the vascular BS appears surrounded by one layer of sheath cells, with chloroplasts in a centrifugal position (Figure 2.2a). No mestome sheath (MS) appears to be present between metaxylem vessel elements and laterally adjacent Kranz cells as in other grasses, e.g. Pooideae (Esau, 1977). According to ultrastructural studies *V. zizanioides* possesses a Kranz anatomy with dimorphic chloroplasts. The bundle sheath chloroplasts form larger and more numerous starch grains than the mesophyll chloroplasts and in contrast to the latter, they show a reduced grana development or none at all. In the BS, ultrastructural observations show a cell wall suberized lamella (Figure 2.2b) along with the presence of agranal chloroplasts. Sometimes these chloroplasts show an apparent distortion of the thylakoid system somewhere in the central part of the stroma (Figure 2.2b–c). Chloroplasts were observed to contain numerous starch grains (Figure 2.2b) and/or a developed peripheral reticulum, a system of anastomosing tubules contiguous with the inner membrane of the chloroplast envelope (Figure 2.2c). A few, small, and heavily cristated mitochondria were also observed. The mesophyll chloroplasts show most of the thylakoids staked in grana, the complete absence of starch grains, and the presence of several plastoglobules (Figure 2.2d–e) (Bertea *et al.*, 2001).

Rubisco immunolocalization

Immunocytochemical studies were conducted by Giaccone *et al.* (1990, 1991) using *V. zizanioides* young and old leaves in order to localize the enzyme Rubisco and give a clear indication of the photosynthetic pathway adopted by this plant.

For these experiments purified rabbit polyclonal antibodies raised against Rubisco were employed. Bound antibodies were then visualized by linking conjungated gold labelled goat anti-rabbit polyclonal antibodies. High resolution immunolocalization with electron microscopy shows that labelling occurred only in the bundle sheath chloroplasts. Gold particles appeared uniformly distributed in the entire organelles with the exception of starch grains and plastoglobules (Figure 2.3). No labelling was observed in the mesophyll chloroplasts (Figure 2.2e) (Bertea *et al.*, 2001).

These studies provided evidence for localization of Rubisco in bundle sheath chloroplast stroma as it was expected for a C_4 plant.

Root anatomy

Vetiveria zizanioides root anatomy, along with leaf anatomy, was initially studied by Prat in 1937. This research, conducted on Andropogoneae from Africa permitted the classification of this species. More recently, Kartusch and Kartusch (1978) have been interested in vetiver root anatomy in order to describe the site of essential oil production.

Other studies related to root anatomy and essential oil secretion were those conducted by Viano *et al.* (1991a, 1991b). These authors analysed the anatomy and the ultrastructure of *V. zizanioides* root cultivated on the Reunion Island. The descriptions

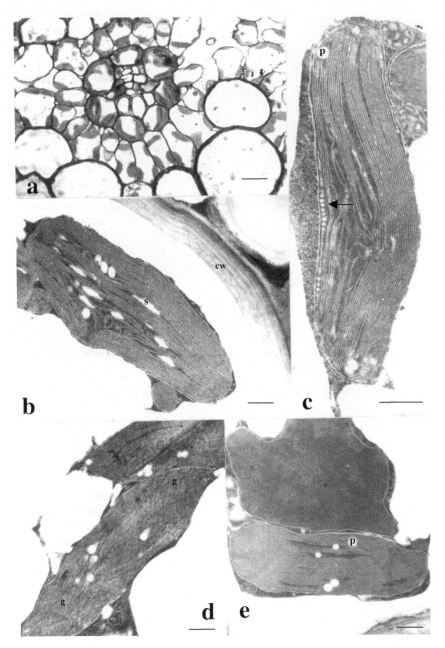

Figure 2.2 a Semi-thin cross section of *V. zizanioides* leaf stained with toluidine blue.
The vascular bundle in a minor vein is surrounded by a layer of sheath cells,
with chloroplasts in a centrifugal position. *b* Bundle sheath chloroplasts
without grana and with an apparent distortion of thylakoids system
somewhere in the central part of the stroma. Note the small starch grains
(S) and the suberized lamella in the BS cell wall (CW). *c* Bundle sheath
chloroplasts, note the peripheral reticulum (arrow) along the chloroplast.
P = plastoglobule. *d–e* Mesophyll chloroplasts with most of the thylakoids
stacked in grana (G) and without starch grains. P = plastoglobule. *a* scale
bar = 4 µm; *b–e* scale bars = 0.5 µm.

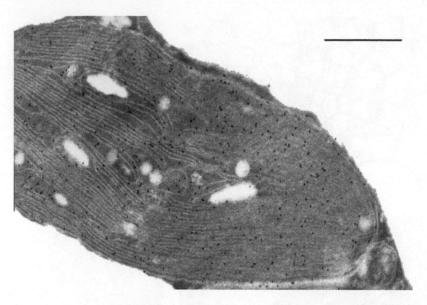

Figure 2.3 High magnification of a bundle sheath chloroplast from *V. zizanioides*, showing immunolabelling against Rubisco. Colloidal gold particles strongly label bundle sheath chloroplast stroma while mesophyll chloroplasts and other organelles were essentially unlabelled. For comparison see Figure 2e. Scale bar = 0.5 µm.

given by these authors are in accordance with our microscopical observations, conducted on *V. zizanioides* Stapf., cultivated at the Botanical Garden of the University of Turin.

V. zizanioides possesses a typical structure of the monocotyledon plants (Figure 2.4). Figure 2.4a shows a scanning electron micrograph of a cross section of *V. zizanioides* root while Figure 2.4b presents a portion of the same root at higher magnification. The transverse section shows a clear separation between the usual three systems, the epidermis, the cortex and the vascular tissue, that form a hollow cylinder due to the presence of the pith. The root cortex, composed of parenchyma cells, is divided in two parts, viz. external (ec) and internal (ic) cortex. The first is composed of 4 or 5 cell layers. In the internal cortex it is possible to observe a tissue similar to an aerenchyma, with intercellular spaces developing into large lacunae of lysigenous origin (LL). Such cortex is common in plants growing in moist habitats but it is also encountered in species growing in drier regions (Esau, 1977). These lacunae, exclusively present in the cortical parenchyma, are separated from each other by a thin cell line. The inner layer of the cortical parenchyma is represented by the endodermis (E). In the absorbing region of the root the endodermal cell wall contains suberin in a bandlike region extending around the cells, especially within the inner tangential and radial walls (casparian strip). In cross section the endodermis cells appear tightly connected. The vascular cylinder or stele comprises the vascular tissues and one or more layers of nonvascular cells, the pericycle (Pe). A circle of metaxylem vessels surrounds a pith composed of cells accumulating starch grains and having thin walls (Figure 2.4c).

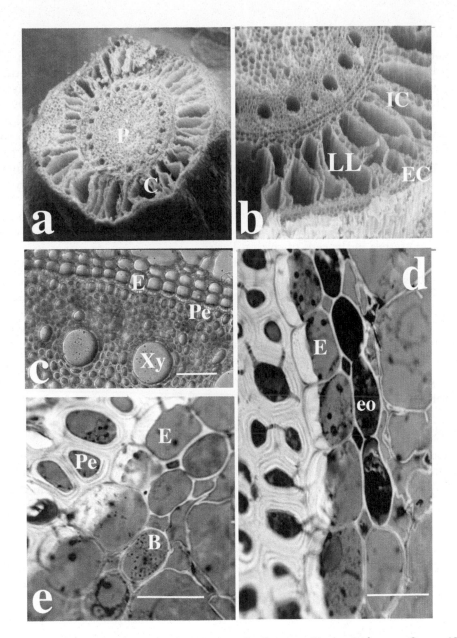

Figure 2.4 a Electron scanning micrograph of mature *V. zizanioides* root. Cortex (C) and the pith (P) are evident. *b* Enlarged view from (*a*) showing the structure at higher magnification (100x). The cortex is composed of two parts: external (EC) and internal (IC) cortex. The latter is occupied by large lysigenous lacunae (LL), characteristic of this species. The vascular tissue surrounds a pith (P) (40x). *c* Transverse section of mature *V. zizanioides* root stained with Congo Red, Methyl Green, and Sudan Black B, showing the thick-walled endodermis (E), the pericycle (Pe) and the vascular tissue (Xy) surrounding the pith composed of parenchymatous cells. Scale bar = 100 μm. *d* Cross section of a *V. zizanioides* root, stained with Sudan Black B, showing oil-producing cells within the last layer of the cortical parenchyma. These cells are close to the endodermis (E) and appear heavily coloured (eo). Scale bar = 30 μm. *e* Enlarged view of the site of essential oil production in the root of *V. zizanioides*. Numerous bacteria (B) appear in the oil-producing cells. The nature of these bacteria is still unknown. Scale bar = 30 μm.

We observed that the secretion of essential oil in *V. zizanioides* occurs within the cortical layer and close to the endodermis. Figure 2.4d shows a cross section of vetiver root, where the essential oil-producing cells are evidenced by treatment with Sudan Black B. Our results are in accordance with those obtained by Viano *et al.* (1991a, 1991b). These authors analysed root ultrastructure using electron transmission microscopy and they detected essential oil crystals in the inner cortical layer close to the endodermis. According to these authors the secretion of the essential oil occurs in this region and successively reaches the whole cortex. The oil density increased as a function of the age of the root, becoming more and more viscous and forming crystal structures in old roots.

They also observed numerous bacteria in the parenchymatous essential oil-producing cells and in the lysigenous lacunae. It seemed that these bacteria coexisted with essential oil crystals in the cells of cortical layer.

Similar results were obtained in our studies. Figure 2.4e shows a light microscopy photograph revealing a portion of the cortical layer region responsible for essential oil production. It is possible to identify numerous bacteria (B) in the parenchymatous cells, where essential oil is produced.

Until now, nothing has been known about the type and function of these bacteria in vetiver root. The determination of the role played by these microorganisms in the essential oil metabolism will become important.

Biochemistry and Physiology

General consideration

Anatomical studies conducted on *V. zizanioides* leaves indicated the presence of the C_4 Kranz anatomy in this plant along with several ultrastructural features typical of the NADP-ME species. Since the determination of the photosynthetic mechanism could be achieved using biochemical approaches, *V. zizanioides* was also studied by analysing the main enzymes involved in this process.

From a biochemical point of view, the three variations on the basic C_4 pathway differ principally in the C_4 acid transported into the bundle sheath cells (malate and aspartate) and in the manner of decarboxylation, and they are named according to the enzymes that catalyse their decarboxylation reactions, NADP-dependent malic enzyme (NADP-ME) found in chloroplasts, NAD-dependent malic enzyme (NAD-ME) found in mitochondria and phosphoenolpyruvate (PEP) carboxykinase (PCK), found in the cytosol of the bundle sheath cells.

The primary carboxylation reaction, common to all three variants, occurs in the cytosol of the mesophyll cells and is catalysed by phosphoenolpyruvate carboxylase (PEP-case), using HCO_3^- rather than CO_2 as a substrate. The fate of the oxaloacetate produced in this reaction depends on the C_4 variant (Gutierrez *et al.*, 1974). In the NADP-ME type, oxaloacetate is reduced to malate in the mesophyll chloroplasts, then transported to the bundle sheath cell chloroplasts and decarboxylated by NADP-ME enzyme. In the NAD-ME and PCK species, oxaloacetate undergoes transamination in the cytosol with glutamate as amino donor. The aspartate is transported into the bundle sheath cells and reconverted to oxaloacetate by transamination in the mitochondria (NAD-ME species) or the cytosol (PCK species). Without changing compartmentalization, the oxaloacetate is reduced and then decarboxylated by NAD-ME in NAD-ME species, while in PCK species oxaloacetate is decarboxylated by PCK. In NADP-ME plants, the C_3 acid transported back to the mesophyll is pyruvic acid as pyruvate but in NAD-ME and PCK species alanine is probably converted to pyruvate in the mesophyll cell cytosol. The final reaction of the C_4 pathway, which is common to all three variants, is the conversion of pyruvate to phosphoenolpyruvate within the mesophyll chloroplasts (Maffei, 1999).

C_4 plants show higher rates of photosynthesis at high light intensities and high temperatures due to the increased efficiency of the PCR cycle (Hatch, 1988). In favourable environments C_4 plants outperform C_3 plants, making them the most productive crops and the worst weeds. Maize, sugarcane and sorghum are examples of C_4 crops.

The genes encoding the C_4 pathway enzymes, including pyruvate, orthophosphate dikinase (PPDK), phosphoenolpyruvate carboxylase (PEP-case), NADP-malate dehydrogenase (NADP-MDH), NAD-malic enzyme (NAD-ME), and phosphoenolpyruvate carboxykinase (PCK), are highly expressed in C_4 plants and, in addition, they are differentially expressed in the two photosynthetic cell types (Sheen and Bogorad, 1987; Langdale et al., 1988; Nelson and Langdale, 1989; Maurino et al., 1997).

One of the differences between C_4 and C_3 photosynthesis concerns the level of expression of the enzyme involved in the C_4-carbon pathway (Gallardo et al., 1995).

A feature of all the enzymes implicated in the C_4 pathway is their high activity in leaves relative to the activity recorded for C_3 plants (15- to >100-fold) (Ashton et al. 1990).

Photosynthetic mechanism: biochemical determination

Preliminary studies on the photosynthetic apparatus provided localization of the key enzyme ribulose-1, 5-bisphosphate carboxylase/oxygenase (Rubisco) exclusively in bundle sheath cells (Giaccone et al., 1990, 1991) as reported before.

In order to better characterize the photosynthetic mechanism of these species $\delta^{13}C$ analyses, some kinetic characteristics of PEP-case (E.C. 4.1.1.31) and the highest activities of Rubisco (E.C. 4.1.1.39) and glycolate oxidase (GO, E.C. 1.1.3.1) were estimated by Maffei et al. (1995). The results of this investigation are reported below.

The $\delta^{13}C$ analyses were conducted according to O'Leary et al. (1992) in three distinct portions of the leaf blade (apical, central and basal) in order to determine the area having the highest activity of the photosynthetic carboxylating enzyme pool. Results indicated a $-14\%o$ value for the apical portion of the leaf blade, whereas for central and basal portions the values were $-13.7\%o$ and $13.2\%o$, respectively.

These values indicated in general a C_4 photosynthetic mechanism (Edwards and Walker, 1983; Korner et al., 1988) which agrees with the low CO_2 compensation point levels (<10 $\mu l \ l^{-1} \ CO_2$) obtained by Krenzer et al., 1975 for this plant. According to Sasakawa et al. (1989) the variation in $\delta^{13}C$ appears, in maize, to be directly correlated with variations in the level of PEP-case. Also, in the case of V. zizanioides, preliminary assays for PEP-case activity resulted in an increasing trend from the base to the tip of leaves.

Table 2.1 shows a direct comparison of photosynthetic enzyme activities from V. zizanioides and some C_3 plants (Brachypodium pinnatum, Mentha piperita, Spinacia oleracea, and Trifolium repens) and C_4 plants (Cymbopogon citratus, Saccharum officinarum, Sorghum bicolor, and Zea mays) growing in the same habitat. Extractions and assays were conducted according to Ashton et al. (1990).

Table 2.1 Photosynthetic enzyme activities (μKat g−1 fresh weight) from *Vetiveria zizanioides* and some C_3 and C_4 plant extracts.

Plants and photosynthetic pathway	*GO	Rubisco	*PEP-case
Vetiveria zizanioides	15.55	57.58	2,830.31
C_3			
Brachipodium pinnatum	n.d.	n.d.	56.66
Mentha piperita	119.00	178.02	41.11
Spinacia oleracea	37.32	313.31	26.70
Trifolium repens	38.41	466.52	53.74
C_4			
Cymbopogon citratus	15.61	77.12	1,527.38
Saccharum officinarum	n.d.	n.d.	2,258.24
Sorghum bicolor	n.d.	n.d.	3,157.01
Zea mays	7.82	52.14	320.00

n.d. = not determined; (standard error of the mean).
* For abbreviations see text

Considering the glycolate oxidase (GO) activity, very high values were obtained for the C_3 plant *M. piperita*, confirming previous data on this plant (Maffei and Codignola, 1990), whereas lower values were found for *T. repens* and *S. oleracea*. Very low GO activities were found for *C. citratus, Z. mays* and *V. zizanioides* indicating, in the latter, low levels of photorespiration as is typical of C_4 plants (Edwards and Walker, 1983; Huang *et al.*, 1983). Rubisco activities were higher in C_3 plants when compared to C_4 plants (Edwards and Walker, 1983), and even in this case *V. zizanioides* showed values typical of C_4 plants. PEP-case activity ranged between 26.70 and 56.66 μKat g^{-1} fresh weight in C_3 plants, whereas much higher values were recorded for C_4 plants with the highest activity found in *S. bicolor*. Also for PEP-case, *V. zizanioides* showed activities typical of C_4 plants (Ting and Osmond, 1973; Vidal *et al.*, 1983; Maffei *et al.*, 1988).

Apparent Michaelis constants (K_m) of PEP-case were estimated for PEP and Mg^{2+} using the desalted preparation. The K_m Mg was estimated from double reciprocal plots 1/v against 1/[S] and was in *V. zizanioides* 0.59 mM whereas for *C. citratus* and *Z. mays* it was 0.78 and 0.70 mM, respectively. K_m PEP and *n* number were estimated from a plot of log v/(V-v) against log [S] according to Ting and Osmond (1973) (Figure 2.5) and were 0.39 mM and 1.52, respectively, for *V. zizanioides*. For *C. citratus* and *Z mays* K_m and *n* number values were respectively 0.60 mM and 1.02 and 0.53 mM and 1.06; for *T. repens* and *B. pinnatum*, 0.31 mM and 0.95 and 0.63 mM and 0.83, respectively.

Temperature and pH variations have a significant effect on photosynthetic enzymes such as PEP-case (Wu and Wedding, 1987; Angelopoulos and Gavalas, 1988; Stiborova, 1988; Ashton *et al.*, 1990; Grahame and Latzko, 1993) since the change of these parameters can promote the dissociation of this enzyme which is a homotetramer with a molecular weight around 400,000 (Iglesias *et al.*, 1986; Angelopoulos and Gavalas, 1988, Ashton *et al.*, 1990). The dissociation of the tetramer in to smaller subunits (less active or inert dimers) is related to low temperatures and high pH (Angelopoulos and Gavalas, 1988). pH changes significantly affected the kinetics of PEP binding in *V. zizanioides* (Figure 2.6a) and desalted extract showed maximum activity at pH 8.0.

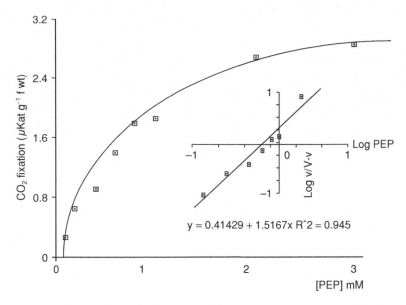

Figure 2.5 Rate curve for phosphoenolpyruvate carboxylase from *V. zizanioides* as a function of phosphoenolpyruvate concentration along with the log-log form for the calculation of *n* number and *Km*.

However, when the assay was conducted at cytoplasmic pH values (7.0–7.5) activity was superior to that detected at pH values greater than 8.3. Temperatures below 25° or above 32 °C were found to inactivate PEP-case in desalted extract and the highest activity was detected at 30 °C (Figure 2.6b).

Characterization of the photosynthetic variant

Further studies dealing with the characterization of the C_4 variant indicated a NADP-dependent malic enzyme photosynthetic pathway in *V. zizanioides*.

The biochemical type was established through the estimation of the highest activities of NADP-dependent malic enzyme (NADP-ME, EC 1.1.1.40), NADP-dependent malate dehydrogenase (NADP-MDH, E.C. 1.1.1.82), pyruvate, orthophosphate dikinase (PPDK, E.C. 2.7.9.1), NAD-dependent malic enzyme (NAD-ME, E.C. 1.1.1.39) and phosphoenolpyruvate carboxykinase (PCK, E.C. 4.1.1.49) and some kinetic parameters of NADP-ME and NADP-MDH (Bertea *et al.*, 2001).

Extraction and partial purification sequentially involved precipitation with crystalline ammonium sulfate, dialysis and anion exchange (DEAE-Sephacell). Both extraction and assays were conducted according to Ashton *et al.* (1990). Table 2.2 shows photosynthetic enzyme activities in DEAE-extracts of *V. zizanioides* Stapf.

PPDK activity was very low, 25.70 nKat mg^{-1} chl. and PCK activity was even lower (<1 nKat g^{-1} mg chl). Similar results were obtained for NAD-ME. However, the activity of PPDK in *V. zizanioides* agrees with the literature data for C_4 plants (Ashton *et al.*, 1990). The extractable activity of PPDK which generates phospoenolpyruvate, the CO_2 acceptor common to all C_4 variants (Grahame and Latzko, 1993) is only just

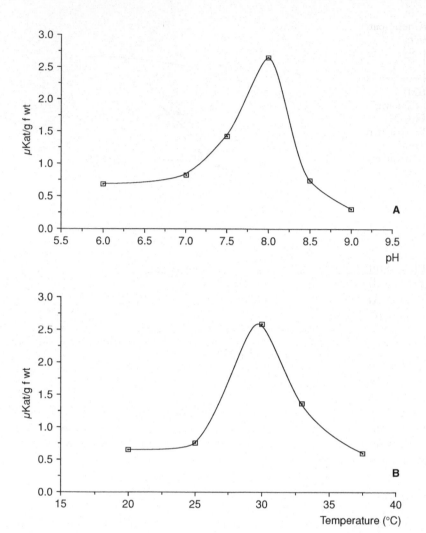

Figure 2.6 Effect of pH *(a)* and temperature *(b)* on PEP-case activity on *V. zizanioides* desalted extracts.

Table 2.2 Photosynthetic enzyme activities from *Vetiveria zizanioides* DEAE-extracts.

*Photosynthetic enzymes	Enzyme activities
PPDK (nKat mg^{-1} chl)	25.70 (±0.34)
NAD-ME (nKat mg^{-1} chl)	35.68 (±0.68)
PCK (nKat mg^{-1} chl)	<1.0 (n.d.)
NADP-MDH (μKat mg^{-1} chl)	5.64 (±0.48)
NADP-ME (μKat mg^{-1} chl)	4.42 (±0.26)

*For abbreviations see text; (standard error of the mean).

Table 2.3 Kinetic parameters of NADP-MDH and NADP-ME from *Vetiveria zizanioides* DEAE-extracts.

Kinetic parameters	NADP-MDH	NADP-ME
K_m OAA (μM)	81 ($\pm$0.340)	–
V_{ma} OAA (μKat mg^{-1} chl)	5.48 ($\pm$0.021)	–
K_m NADPH (μM)	133 ($\pm$0.707)	–
V_{ma} NADPH (μKat mg^{-1} chl)	7.80 ($\pm$0.012)	–
K_m Malate (mM)	–	3.3 ($\pm$0.721)
V_{max} Malate (μKat mg^{-1} chl)	–	1.5 ($\pm$0.507)
K_m NADP$^+$ (μM)	–	102 ($\pm$24.04)
V_{ma} NADP$^+$ (μKat mg^{-1} chl)	–	0.026 ($\pm$0.007)

(standard error of the mean).

sufficient to account for observed rates of photosynthesis, suggesting that this enzyme is potentially rate limiting (Furbank and Taylor, 1995).

The highest NADP-MDH and NADP-ME activities were therefore measured and gave a clearer picture of the *V. zizanioides* C$_4$ photosynthetic pathway.

Kinetic characteristics (K_m and V_{max}) of NADP-MDH and NADP-ME were comparable to those of plants belonging to NADP-dependent malic enzyme photosynthetic variant. The apparent Michaelis constant for NADP-MDH was calculated for oxaloacetic acid (OAA) and NADPH, while for NADP-ME K_m was measured for malate and NADP$^+$, using DEAE preparations (Table 2.3) and estimated from a plot according to Lineweaver-Burk. K_m values obtained for NADPH and OAA were 0.133 mM and 0.081 mM respectively.

The K_m values for NADP$^+$ and malate were 0.102 mM and 3.30 mM, respectively. The saturation curves obtained when the velocity of reaction was measured as a function of NADPH and OAA (NADP-MDH) and NADP$^+$ and malate (NADP-ME) were typically hyperbolic. Figure 2.7 shows curves obtained with OAA (NADP-MDH) (Figure 2.7a) and malate (NADP-ME) (Figure 2.7b). In general, the high enzyme activities of NADP-MDH and NADP-ME allowed the characterization of the enzymes and yielded the evidence for an NADP-ME variant in *V. zizanioides*. The kinetic properties of both enzymes are consistent with a high photosynthetic activity, even if the plants were cultivated in temperate climates.

NADP$^+$ *inhibition of NADP-MDH*

In *Zea mays*, activation of NADP-MDH is regulated by oxidation and reduction of cystein residues (thioredoxin-mediated system) (Lunn *et al.*, 1995) and interconversion of the reduced and oxidized forms is influenced by the NADPH/NADP$^+$ ratio (Trevanion *et al.*, 1997). A high NADPH/NADP ratio leads to more active enzymes; thus high rates of OAA reduction only occur in reduced conditions (Ashton *et al.*, 1990). These properties of the enzyme provide it with a sensitive mechanism by which activity can respond to changes in the redox potential of the chloroplast (Edwards *et al.*, 1985). A high ratio leads to more active enzymes; thus high rates of OAA reduction only occur in reduced conditions. The percentage of inhibition caused by the increase of NADP$^+$ concentration in our DEAE-extracts was in accordance with these observations.

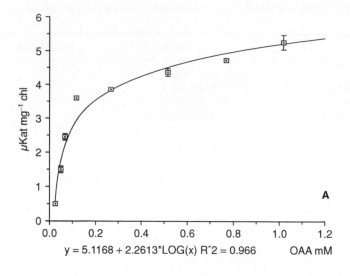

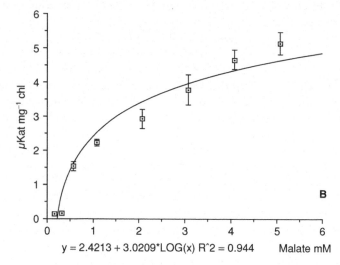

Figure 2.7 a Rate curve for NADP-MDH from *Vetiveria zizanioides* as a function of oxaloacetic acid (OAA) concentration. *b* Rate curve for NADP-ME from *V. zizanioides* as a function of malic acid concentration.

The enzyme assayed at saturation concentration of NADPH and OAA and in the presence of different $NADP^+$ amounts showed an inhibition response. The inhibition percentage increased with $NADP^+$ concentrations. The effect at low concentrations was not very high, but the activity recorded in the presence of 0.25 mM $NADP^+$ was only the 30% of the activity measured in the absence of the oxidized coenzyme (Figure 2.8).

pH and temperature dependence of NADPH-MDH and NADP-ME

Climatic conditions exert an evident effect on the physiological status of the photosynthetic enzymes, and variations in light, temperature, moisture, etc. may influence

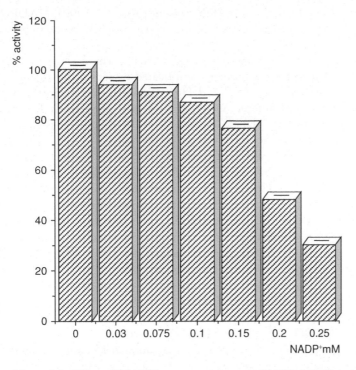

Figure 2.8 Effect of NADP⁺ concentrations on NADP-MDH activity of *Vetiveria zizanioides* DEAE extracts

the cytosolic and stromal pH. As the metabolism of plants is affected by changes in environmental temperatures it is important to evaluate enzyme activities at different temperature values. The response to these changes depends on the photosynthetic pathway adopted, resulting in a different optimum range of temperature over which the highest growth rate can be maintained (Fitter and Hay, 1987). In order to evaluate an eventual adaptation of *V. zizanioides*, photosynthetic enzyme activities were recorded at the low and high temperature values typical of the humid-temperate climates.

Activities were also measured at different pH values in order to estimate increase or decrease in the reaction rate due to changes of protein conformation and consequent pH fluctuations.

In C_4 plants, NADP-MDH has subunits of 42 kDa and the native enzyme apparently occurs as either a tetramer or a dimer (Ashton *et al.*, 1990). The tetramer is the more active form and is stable at alkaline pH levels and at high temperatures (Ashton *et al.*, 1990).

With regard to NADP-ME, the polymerization state of this photosynthetic enzyme changes noticeably in the physiological pH range of 7.0 to 8.0 (Grahame and Latzko, 1993). NADP-ME is a tetramer with a molecular weight around 280 kDa (Ashton *et al.*, 1990) and is more stable at pH values greater than 8.0 (Edwards and Andreo, 1992). This enzyme exists as a dimer and a monomer, both of which are active. The NADP-ME from sugar cane leaves may be a monomer, dimer or tetramer depending on the pH level. At pH 8.0 it is predominantly a tetramer, the most active form, while at pH 7.0 the dimer is the main form (Edwards and Andreo, 1992). NADP-ME purified from green leaves of maize is also a tetramer constituted by monomers of 62 kDa (Maurino *et al.*, 1997).

The differences in the pH values can dramatically change the activities of these photosynthetic enzymes. The pH effect may be complicated by sensitivity to contaminating heavy metal ions such as Cd^{2+}, Zn^{2+}, and Hg^{2+} (Edwards and Andreo, 1992).

Temperature is also a critical parameter. When it is low, the photosynthesis rates in C_4 plants may fall below that of C_3 plants due to a decrease of activity as the result of dissociation from active to non-active forms (Angelopoulos and Gavalas, 1988). Because they originated in tropical and subtropical areas, the optimum temperature for photosynthesis in C_4 plants is 30–40 °C, which is approximately 10 °C higher than in C_3 plants (Leegood, 1993). However, C_4 photosynthesis is usually sensitive to low temperatures; the minimum temperature for photosynthesis in several C_4 tropical grasses is 5–10 °C. The inhibition of photosynthesis under low temperature may be related to damage to enzymes, pigment-protein complexes or membranes essential to the normal functioning of photosynthetic carbon metabolism (Matsuba *et al.*, 1997). The response of enzyme activities to such changes depends on the photosynthetic pathway adopted, resulting in a different optimum range of temperatures over which the highest growth rate can be maintained (Fitter and Hay, 1987).

Figure 2.9 shows the results obtained with the assays at different pH and temperature values using NADP-MDH and NADP-ME DEAE extracts.

The maximum activity of NADP-MDH enzyme was found at pH 8.5 in accordance with the high activity at alkaline pH values. At pH 7.0–7.5 the activity was comparable to that at pH values >8.5 (Figure 2.9a). A constant activity increase was observed starting from the lowest pH value (6.0) up to pH 8.5. Temperature changes also affected the reaction rate (Figure 2.9b). When the enzyme activity was measured using the standard assay systems at temperatures ranging from 20 to 49 °C, the maximum activity was detected at 45 °C, while at 49 °C activity was lower, but still higher than that at 20 °C in accordance with the typical behaviour of C_4 photosynthetic enzymes.

Figure 2.10 depicts the NADP-ME activities at different pH and temperature values. The highest activity in *V. zizanioides* was recorded at pH 8.3 in accordance with the enzyme characteristics, whereas at pH 10.5 the activity was higher than that recorded at pH 6.0 and 6.5 (Figure 2.10a). With regard to the temperatures the maximum activity was measured at 45 °C, the lowest at 20 °C (Figure 2.10b). Also in the latter case activity responses to temperature changes are typical of C_4 plants (Casati *et al.*, 1997).

Chemical-physical parameters (activities at different temperatures and pH values) of *V. zizanioides* NADP-MDH and NADP-ME indicated that this species is a C_4 NADP-ME plant which is able to retain its photosynthetic mechanism even when cultivated in temperate climates.

CO$_2$ assimilation and stomatal conductance measurements

The dissipative effects of photorespiration are avoided in C_4 plants by the mechanism that concentrate CO_2 at the carboxylation sites in the bundle sheath chloroplasts. By expressing the photosynthetic rates as a function of CO_2 concentrations it is possible to calculate the compensation point (the CO_2 concentration at which CO_2 assimilation is zero). The remarkable differences between the photosynthetic responses of C_3 and C_4

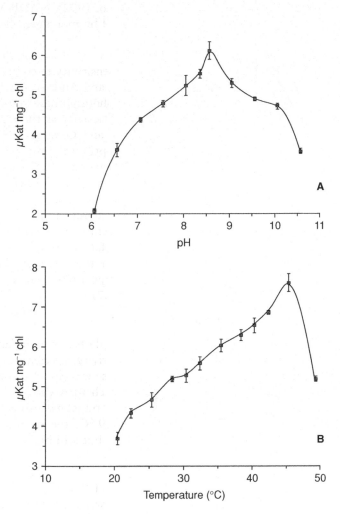

Figure 2.9 Effect of pH *(a)* and temperature *(b)* variation on NADP-MDH activity of *Vetiveria zizanioides* DEAE extracts.

plants to CO_2 became apparent in this type of analysis. In plants with CO_2 concentrating mechanisms, including C_4 plants, the CO_2 concentrations at the carboxylation sites are often saturating. Plants with C_4 metabolism have a CO_2 compensation point of zero or nearly zero, reflecting their very low levels of photorespiration (Maffei, 1999).

In addition, the C_4 mechanism allows the plant to maintain high photosynthetic rates at lower partial pressure CO_2 values in the intercellular spaces of the leaf, which require lower rates of stomatal conductance for a given rate of photosynthesis (Maffei, 1999).

For the reasons described above the measurements of CO_2 compensation point and stomatal conductance can be useful for distinguishing between C_3 and C_4 pathways.

Previous work done by Krenzer *et al.* (1975) indicated a low CO_2 compensation point suggesting a C_4 mechanism for *V. zizanioides*, even if low CO_2 compensation points levels are also characteristic of C_3-C_4 intermediate plants (Maffei *et al.*, 1988).

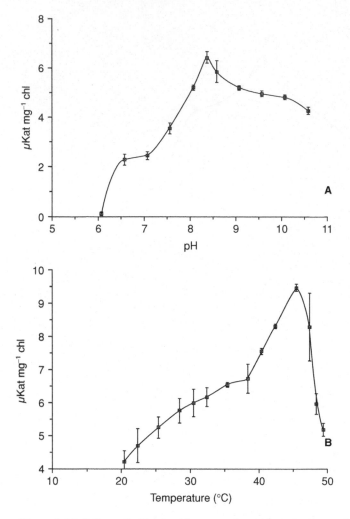

Figure 2.10 Effect of pH (*a*) and temperature (*b*) variation on NADP-ME activity of *Vetiveria zizanioides* DEAE extracts.

This parameter along with stomatal conductance was evaluated on *V. zizanioides* in our laboratory using a CIRAS Infra Red GAS Analyzer equipped with a Parkinson leaf cuvette (PP System) (Bertea *et al.*, 2001). Figure 2.11a shows the changes in photosynthetic rates as a function of CO_2 concentrations. *V. zizanioides* possesses a very low compensation point (between 8 and 22 ppm CO_2) as was expected for a C_4 plant. Figure 2.11b depicts the stomatal conductance as a function of CO_2 concentrations. Stomatal aperture increases in response to CO_2 concentrations up to 157 pmm. However, at higher CO_2 values a decrease was recorded, thus indicating a clear effect of the CO_2 concentrating mechanism present in C_4 plants. This process allows the leaf to maintain high photosynthetic rates at lower CO_2 intercellular concentrations, which require lower rates of stomatal conductance.

CO_2 assimilation and stomatal conductance were also measured on primordial, young, mature and old *V. zizanioides* leaves in order to evaluate the changes in photosynthesis

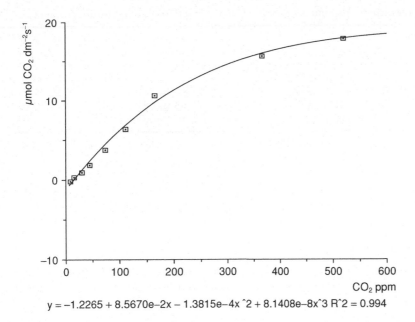

$$y = -1.2265 + 8.5670e{-}2x - 1.3815e{-}4x\char`^2 + 8.1408e{-}8x\char`^3 \quad R\char`^2 = 0.994$$

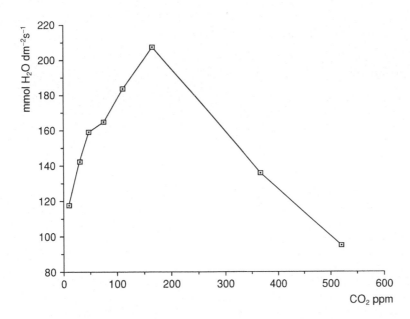

Figure 2.11 Changes in photosynthesis (*a*) and in stomatal conductance (*b*) as a function of environmental CO_2 concentrations.

as a function of the leaf age. The results are reported in Table 2.4. Primordial leaf presented the lowest CO_2 assimilation value (7.17 μmol CO_2 dm^{-2} s^{-1}), while the highest values were recorded in young and mature leaves, without appreciable differences (12.82 and 12.83 μmol CO_2 dm^{-2} s^{-1}, respectively). The CO_2 assimilation value in old leaf was still high (11.30 μmol CO_2 dm^{-2} s^{-1}).

Table 2.4 Photosynthetic parameters measured in *V. zizanioides* leaves of different ages.

Leaf age	PAR	*CO$_2$ assimilation	*Stomatal conductance
primordial	1,055.6 (155.80)	7.17 (0.22)	192.90 (2.85)
young	806.9 (132.20)	12.82 (0.10)	201.85 (0.13)
mature	1,370.7 (46.15)	12.83 (0.08)	223.70 (1.10)
old	1,162 (50.59)	11.30 (0.08)	96.75 (0.52)

*CO$_2$ assimilation is expressed as μmol CO$_2$ dm^{-2} s^{-1}, stomatal conductance as mmol H$_2$O dm^{-2} s^{-1} (standard error of the mean).

With regard to stomatal conductance, the lowest value was measured in old leaf (96.75 H$_2$O dm^{-2} s^{-1}), while the highest occurred in mature leaf (223.70 H$_2$O dm^{-2} s^{-1}).

Concluding Remarks

The results described above indicate that *V. zizanioides* is an NADP-ME type C$_4$ plant. The combination of economic (production of essential oils), ecological (prevention of soil erosion) and symbiotic properties (endobacteria) and the ease with which callus induction and plant regeneration can be obtained from leaf explants (Mucciarelli *et al.*, 1993), as well as its high C$_4$ photosynthetic efficiency and the ability to retain a high enzyme activity, even when cultivated in temperate climates, make this plant an interesting object for future applications.

Acknowledgements

The authors are grateful to Mr. Giovanni D'Agostino for providing photographs of leaf ultrastructure and immunolocalization of Rubisco and to Dr. Marco Mucciarelli for the photographs related to root anatomy.

References

Akhila, A. and Thakur, R.S. (1989) Biosynthesis of the constituent of vetiver oil II. Nootkatane and eudesmane compounds. *11th Int. Cong. Essentl Oils, Frag. Flav.*, New Delhi, November 1989.

Angelopoulos, K. and Gavalas, N.A. (1988) Reversible cold inactivation of C$_4$ phosphoenol-pyruvate carboxylase: factors affecting reactivation and stability. *Journal of Plant Physiology* **132**, 714–719.

Ashton, A.R., Burnell, J.N., Furbank, R.T., Jenkins, C.D.L. and Hatch, M.D. (1990) Enzymes of C$_4$ photosynthesis. In P.M. Dey and J.B. Harborne (eds.), *Methods in Plant Biochemistry*, Academic Press, London, pp. 39–72.

Bertea, C.M., Scannerini, S., D'agostino, G., Mucciarelli, M., Camusso, W., Bossi, S., Buffa, G. and Maffei, M. (2001) Evidence for a C$_4$ NADP-ME photosynthetic pathway in *Vetiveria zizanioides* Stapf. *Plant Biosystems* (in press).

Casati, P., Spampinato, C.P. and Andreo, C.S. (1997) Characteristics and physiological function of NADP-malic enzyme from wheat. *Plant Cell Physiology*, **38**, 928–934.

Chapman, K.S.R. and Hatch, M.D. (1983) Intracellular location of phosphoenolpyruvate carboxy-ikinase and other C$_4$ photosynthetic enzymes in mesophyll and bundle sheath protoplasts of *Panicum maximum*. *Plant Science Letters*, **29**, 145–154.

Cramer, W.A., Furbacher, P.N., Szczepaniak, A. and Tae, G-S. (1991) Electron transport between photosystem II and photosystem I. *Current Topics in Bioenergetics*, 16, 179–222.

Dalton, P.A., Smith, R.J. and Troung, P.N.V. (1996) Vetiver grass hedges for erosion control on cropped flood plain-hedge hydraulics. *Agricultural Water Management*, 31, 91–104.

Eastman, P.A.K., Dengler, N.G. and Peterson, C.A. (1988) Suberized bundle sheaths in grasses (*Poaceae*) of different photosynthetic types I. Anatomy, ultrastructure and histochemistry. *Protoplasma*, 142, 92–111.

Edwards, G.E. and Andreo, C.S. (1992) NADP-Malic enzyme from plants. *Phytochemistry*, 31, 1845–1857.

Edwards, G.E. and Huber, S.C. (1981) The C_4 pathway. In M.D. Hatch and N.K. Boardman (eds.), *The Biochemistry of Plants. A comprehensive Treatise*, Academic Press, New York, pp. 237–281.

Edwards, G.E., Nakamoto, H., Burnell, J.N. and Hatch, M.D. (1985) Pyruvate, Pi dikinase and NADP-malate dehydrogenase in C_4 photosynthesis: properties and mechanism of light/dark regulation. *Annual Review of Plant Physiology*, 36: 255–286.

Edwards, G.E. and Walker, O.A. (1983) C_3, C_4: *Mechanism, and cellular and environmental regulation, of photosynthesis*, Blackwell Scientific Publications, London.

Esau, K. (1977) *Anatomy of Seed Plants*, 2nd Edition, Wiley & Sons, New York.

Fitter, A.H. and Hay, R.K.M. (1987) *Environmental physiology of plants*, Academic Press, London.

Furbank, R.T. and Taylor, W.C. (1995) Regulation of photosynthesis in C_3 and C_4 plants: a molecular approach. *The Plant Cell*, 7, 797–807.

Gallardo, F., Miginiac-Maslow, M., Sangwan, R.S., Decottignies, P., Keryer, E., Dubois, F., Bismuth, E., Galvez, S., Sangwan-Norreel, B., Gadal, P. and Crétin, P. (1995) Monocotyledonous C_4 NADP$^+$-malate dehydrogenase is efficiently synthesized, targeted to chloroplasts and processed to an active form in transgenic plants of the C_3 dicotyledon tobacco. *Planta*, 97, 324–332.

Giaccone, P., Berta, G. and Maffei, M. (1990) Studi preliminari sul meccanismo fotosintetico di *Vetiveria zizanioides*, una graminacea essenziera. *Giornale Botanico Italiano*, 124, 189.

Giaccone, P., D'Agostino, G. and Luisoni, E. (1991) Immunogold localization of ribulose-1, 5-bisphosphate carboxylase (Rubisco) in chloroplasts of *V. zizanioides*. *European Journal of Basic and Applied Histochemistry*, 35, 51.

Grahame, J.K. and Latzko, E. (1993) Photosynthesis: Carbon Metabolism. Twenty years of following carbon cycles in photosynthetic cells. In: *Progress and Botany* (Behnke, H.D., Lüttge, U., Esser, K., Kadereit, K.E. and Runge, M. eds.), Vol. 54, pp. 174–199, Springer-Verlag, Berlin.

Gutierrez, M., Gracen, V.E. and Edwards, G.E. (1974) Biochemical and cytological relationships in C_4 plants. *Planta*, 119, 279–300.

Hatch, M.D. (1987) C_4 photosynthesis: a unique blend of modified biochemistry, anatomy and ultrastructure. *Biochimica et Biophysica Acta*, 895, 81–106.

Hatch, M.D. (1988) C_4 photosynthesis: a unique blend of modified biochemistry, anatomy and ultrastructure. *Biochimica et Biophysica Acta*, 895, 81–106.

Hatch, M.D. (1992) C_4 photosynthesis: an unlikely process full of surprises. *Plant and Cell Physiology*, 33, 333–342.

Hatch, M.D., Kagawa, T. and Craig, S. (1975) Subdivision of C_4-pathway species based on differing C_4 acid decarboxylating systems and ultrastructural features. *Australian Journal of Plant Physiology*, 2, 111–128.

Hattersley, P.W. and Watson, L. (1976) C_4 grasses: an anatomical criterion for distinguishing between NADP-malic enzyme species and PCK or NAD-malic enzyme species. *Australian Journal of Botany*, 24, 297–308.

Huang, A.H.C., Trelease, R.N. and Moore Jr, T.S. (1983) *Plant peroxisomes*, Academic Press, London.

Iglesias, A.A., Gonzalez, D.H. and Andreo, C.S. (1986) Purification and molecular and kinetic properties of phosphoenolpyruvate carboxylase from *Amaranthus viridis* L. leaves. *Planta*, 68, 239–244.

Jenkins, C.L.D., Furbank, R.T. and Hatch, M.D. (1989) Mechanism of C_4 photosynthesis. *Plant Physiology*, **91**, 1372–1381.

Kartusch, R. and Kartusch, B. (1978) Nachweis und Lokalisierung der wurzel von *Vetiveria zizanioides* L. *Mikroskopie*, **34**, 195–201.

Korner, C., Farquhar, G.D. and Roksandic, Z. (1988) A global survey of carbon isotope discrimination in plants from high altitude. *Oecologia*, **74**, 623–632.

Krenzer, E.G., Moss, D.N. and Crookston, R.K. (1975) Carbon dioxide compensation points of flowering plants. *Plant Physiology*, **56**, 194–206.

Laetsch, W.M. (1974) The C_4 syndrome: a structural analysis. *Annual Review of Plant Physiology*, **25**, 27–52.

Langdale, J.A., Rothermel, B.A. and Nelson, T. (1988) Cellular pattern of photosynthetic gene expression in developing maize leaves. *Genes and Development*, **2**, 106–115.

Leegood, R.C. (1993) Carbon dioxide-concentrating mechanisms. In P.J. Lea and R.C. Leegood (eds.), *Plant Biochemistry and Molecular Biology*, Wiley and Sons, UK, pp. 47–72.

Lemberg, S. and Hale, R.B. (1978) Vetiver oils of different geographical origins. *Perfumer & Flavorist*, **3**, 23–27.

Lunn, J.E., Agostino, A. and Hatch, M.D. (1995) Regulation of NADP-malate dehydrogenase in C_4 plants – activity and properties of maize thioredoxin M and the significance of non-active site thiol groups. *Australian Journal of Plant Physiology*, **22**, 577–584.

Maffei, M. (1999) *Biochimica Vegetale*, Piccin, Padova.

Maffei, M., Codignola, A. and Fieschi, M. (1988) Photosynthetic Enzyme Activities in Lemongrass Cultivated in Temperate Climates. *Biochemical Systematics and Ecology*, **16**, 263–264.

Maffei, M. and Codignola, A. (1990) Photosynthesis, photorespiration and herbicide effect on terpene production in peppermint (*Mentha piperita* L.). *Journal of Essential Oil Research*, **2**, 275–286.

Maffei, M., Scannerini, S., Berta, G. and Mucciarelli, M. (1995) Photosynthetic enzyme activities in *Vetiveria zizanioides* cultivated in temperate climates. *Biochemical Systematics and Ecology*, **23**, 27–32.

Matsuba, K., Imaizumi, N., Kaneko, S., Samejima, M. and Ohsugi, R. (1997) Photosynthetic responses to temperature of phosphoenolpyruvate carboxykinase type C_4 species differing in cold sensitivity. *Plant, Cell and Environment*, **20**, 268–274.

Maurino, V.G., Drincovich, M.F., Casati, P., Andreo, C.S., Edwards, G.E., Ku, M.S.B., Gupta, S.K. and Franceschi, V.R. (1997) NADP-malic enzyme: immunolocalization in different tissues of the C_4 plant maize and the C_3 plant wheat. *Journal of Experimental Botany*, **48**, 799–811.

Metcalfe, C.R. (1960) *Anatomy of the Monocotyledons. I. Gramineae*, Oxford University Press, London.

Mucciarelli, M., Bertea, C.M., Cozzo, M., Gallino, M. and Scannerini, S. (1997) *Vetiveria zizanioides* as a tool for environmental engineering. *Acta Horticolturae*, **457**, 261–269.

Mucciarelli, M., Gallino, M., Scannerini, S. and Maffei, M. (1993) Callus induction and plant regeneration in *Vetiveria zizanioides*. *Plant Cell Tissue and Organ Cultures*, **35**, 267–271.

National Research Council (1993) *Vetiver grass: a thin line against soil erosion*, National Academy Press, Washington.

Nelson, T. and Langdale, J.A. (1989) Patterns of leaf development in C_4 plants. *The Plant Cell*, **1**, 3–13.

O'Leary, M.H., Madhavan, S. and Paneth, P. (1992) Physical and chemical basis of carbon isotope fractionation in plants. *Plant, Cell & Environment*, **15**, 1099–1104.

Prat, H. (1936) Caractères anatomique et histologiques de quelques Andropogonées de l'Afrique occidentale. *Ann. Mus. Colon. Marseille*, Ser. 5, **5**, 25–28.

Prat, H. (1936) La Systématique des Graminées. *Ann. Sci.nat.Bot.*, Ser. 10, **18**, 165–258.

Rodriguez, O.S. (1997) Hedgerows and mulch as soil conservation measures evaluated under field simulated rainfall. *Soil Technology*, **11**, 79–93.

Rothermel, B.A. and Nelson, T. (1989) Primary structure of the maize NADP-dependent malic enzyme. *Journal of Biological Chemistry*, 264, 19857–19592.

Sacco, T. (1960) Possibilità di incremento della *Vetiveria zizanioides* Stapf. Della Somalia, come pianta essenziera. *Rivista di Agricoltura Tropicale e Subtropicale*, 54, 81–87.

Sasakawa, H., Sugiharto, B., O'Leary, M.H. and Sugiyama, T. (1989) $\delta^{13}C$ values in maize leaf correlate with phosphoenolpyruvate carboxylase levels. *Plant Physiology*, 90, 582–585.

Sethi, K.L. and Gupta, R. (1960) Breeding for high essential oil content in Khas (*Vetiveria zizanioides*) roots. *Indian Perfumer*, 24, 72–78.

Sheen, J. and Bogorad, L. (1987) Regulation of levels of nuclear transcripts for C_4 photosynthesis in bundle sheath and mesophyll cells of maize leaves. *Plant Molecular Biology*, 8, 227–238.

Smadia, J., Gaydou, E.M., Lamaty, G. and Conan, J.Y. (1986) Huile essentielle de vétyver Bourbon: influence régionale et climatique sur les constantes physico-chimiques. *Parfums, cosmétiques, arômes*, 69, 69–73.

Smadia, J., Gaydou, E.M., Lamaty, G. and Conan, J.Y. (1988) Essais d'identification des constituants de l'huile essentielle de vétyver Bourbon. *Parfums, cosmétiques, arômes*, 84, 61–66.

Smith, B.N. and Brown, W.V. (1973) The kranz syndrome in the Gramineae as indicated by carbon isotope ratios. *American Journal of Botany*, 60, 505–513.

Stiborova, M. (1988) Phosphoenolpyruvate carboxylase: the key enzyme of C_4-photosynthesis. *Photosynthetica*, 22, 240–263.

Ting, I.P. and Osmond, C.B. (1973) Photosynthetic phospho*enol*pyruvate carboxylase. Characteristics of alloenzymes from leaves of C_3 and C_4 plants. *Plant Physiology*, 51, 439–447.

Trevanion, S.J., Furbank, R.T. and Ashton, A.R. (1997) NADP-Malate dehydrogenase in the C_4 plant *Flaveria bidentis. Plant Physiology*, 113, 1153–1165.

Tscherning, K., Leihner, D.E., Hilger, T.H., Mullersamann, K.M. and Elsharkawy, M.A. (1995) Grass barriers in cassava hillside cultivation – rooting patterns and root growth dynamics. *Field Crop Research*, 43, 131–140.

Viano, J., Gaydou, E.M. and Smadja, J. (1991a) Sur la présence de bactéries intracellulaires dans les racines du *Vetiveria zizanioides* (L.) Stapf. *Rev. Cytol. Biol. Vég. – Bot.*, 14, 65–70.

Viano, J., Smadja, J., Conan, J.Y. and Gaydou, E.M. (1991b) Ultrastructure des racines de *Vetiveria zizanioides* (L.) Stapf (Gramineae). Bull. Mus. Nat. Hist. Nat., Paris, 4me série, section B, *Adansonia*, 13, 61–69.

Vickery, J.W. (1935) The leaf anatomy and vegetative characters of the indigenous grasses of N.S. Wales. *Proceedings of the Linnean Society, N.S.W.*, 60, 340–373.

Vidal, J., Crétin, C. and Gadal, P. (1985) The mechanism of photocontrol of phosphoenolpyruvate carboxylase in sorghum leaves. *Physiologie Végetale*, 21, 977–986.

Watson, L. and Dallwitz, M.J. (1992 onwards) *Grass Genera of the World: Descriptions, Illustrations, Identification, Information Retrieval; including Synonyms, Morphology, Anatomy, Physiology, Phytochemistry, Cytology, Classification, Pathogens, World and Local Distribution, and References.* http://biodiversity.uno.edu/delta/. Version: 18th August 1999. Dallwitz (1980), Dallwitz, Paine and Zucker (1993 onwards, 1998), and Watson and Dallwitz (1994), and Watson, Dallwitz, and Johnston (1986).

Wu, M.-X. and Wedding, R.T. (1987) Temperature effects on phosphoenolpyruvate carboxylase from CAM and C_4 plants. A comparative study. *Plant Physiology*, 85, 497–501.

3 Collection, Harvesting, Processing, Alternative Uses and Production of Essential Oil

Claudio Zarotti

Viale Teodorico, 2 20149 Milan, ITALY

Introduction

Vetiver grass, in particular the species *Vetiveria zizanioides*, has been known to be a useful plant for thousands of years. It is mentioned in the ancient Sanskrit writings and it is part of Hindu mythology. Rural people have used it for centuries for the oil from its roots, for the roots themselves, and for the leaves. Its domestication appears to be in southern India and it has been spread around the world through its value as a producer of an aromatic oil for the perfume industry. In the late part of the last century and in this century the sugar industry, particularly in the West Indies, the off-shore eastern African islands such as Mauritius and Reunion, and Fiji, have used the grass for its soil conservation properties.

Uses and Economic Importance of Vetiver

For centuries vetiver has been used in India both as an aromatic plant and for medicinal purposes, and as a plant used for soil conservation. The scented roots are used directly in the making of mats, baskets, fans, bags, curtains, etc., or indirectly by extraction for the distillation of the essential oil. From India the vetiver spread throughout the Tropics. One particular impetus for the spreading of the plant proved to be the Colonial Period, during which it spread both as an aromatic plant and as a hedge plant. After the Second World War and the subsequent end of colonialism, vetiver declined in importance in many countries.

Erosion control

Recently, many projects have been launched with the aim of increasing the use of vetiver in erosion control. Given its morphological, physiological, and ecological characteristics, as discussed in the previous chapters, it is particularly suited to the formation of hedges with a deep root system. In these countries vetiver is used to slow the run-off of the torrential rains (monsoons) and to slow and stop topsoil erosion, but only in the last decade have such farming practices been seriously considered to the point of study and a clearer definition of both the botanical and agronomic characteristics of the plant, and the technical aspects concerning its planting and cultivation. In this way not only can large enterprises with construction projects on a vast scale make use of vetiver as a plant to control soil erosion, but also, and most importantly,

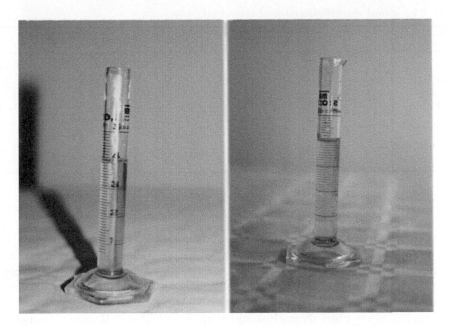

Figure 3.1 Vetiver essential oil. The oil is characterized by its yellow colour, the exact tone of which depends on the roots used, ranging from greenish to reddish. [This figure is also reproduced in the colour section].

individual farmers who, with their own business, have to fight the process of erosion which reduces the fertility of the plots and removes soil and nutrients.

Production of essential oil

In the East, the roots of vetiver have been known for centuries for their scent which is light and pleasant. Vetiver's essential oil, both in its raw form and in other derivative forms, is an important component in luxury perfumes, and therefore in the industry concerned, owing to the delicacy of the aroma and to the amber-scented tones.

Once uprooted the root can immediately be submitted to the distillation process after having been dried naturally, cleaned and cut up.

Vetiver's essential oil is in the form of a viscous liquid which tends to thicken over time. The oil is characterized by its yellow colour, the exact tint of which depends on the roots used, ranging from greenish to reddish (Figure 3.1).

If the original material is taken from young plants the scent has an earthy tone and the colour of the oil is basic green. This is not the case in older plants which are at least two years old.

The yield by distillation carried out starting with the dried roots, varies from 0.5% to 2% in weight, according to the area of origin of the material and the productivity of the distillation equipment.

The best and most sought-after oils are from Giava and from Bourbon (or Réunion); there are however other types of oil, for instance that from Haiti which is in constant productive and commercial development and oils from Brasil, India, and Africa.

Table 3.1 The annual market allotment of vetiver oil.

COUNTRY	PERCENTAGE
U.S.A.	40
France	20
Switzerland	12
England	10
Japan	4
Germany	2.4
The Netherlands	2
Countries of vetiver oil origin	12–16

The world market in vetiver oil at the end of the 1990s was approximately 400 tons and the European Union tax-free price is around 110 ECU/kg.

The average return per hectare of oil can be considered to be around 40 kg/hectare with a gross production equal to around 4,000 ECU/ha.

The annual market allotment of vetiver oil is reported in Table 3.1.

With regard to the roots themselves which are not used for oil extraction and the aerial portions, there is no world market, since such products are utilized and transformed exclusively in the countries of origin. For the transformation to commercial oil local manpower is widely employed, as this process requires manual and artisan procedures. The bleached products derived from vetiver are perfect for local handicraft use.

Other uses

There are several other uses of vetiver in the regions where the plant originates, which can be of economic importance in the Mediterranean areas.

The use of vetiver in the prevention of toxic contamination of water sources

Nitrate from fertilization, heavy metals, and other toxic materials from weed killers and pesticide spraying, once washed into water sources, will cause environmental pollution and contamination.

Studies conducted in a few parts of the world show evidence that vetiver hedgerows planted across slopes can reduce the rate of surface soil loss on sloping land and, at the same time, the deep and dense vetiver root system functions as a barrier, filtering soil debris and toxic substances in water, thus not allowing them to enter ground water strata (Figure 3.2).

Animal feed

The Indian practice of using the vetiver leaves as foodstuff has been tested in the Department of Livestock Development in Thailand.

The trials were conducted on 10 ecotypes of vetiver with the result that Kamphaengpetch 2 provided better nutritive values than other cultivars in terms of

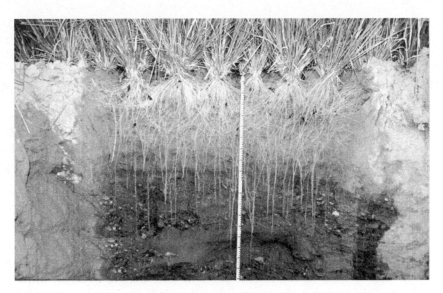

Figure 3.2 Vetiver, owing to its well developed root system, is able to filter contaminant elements such as nitrate residues derived from fertilization, heavy metals and toxic compounds derived from pesticides and other products.

Figure 3.3 Vetiver can be also used for animal feeding.

quantity of total protein, digestible dry matter and minerals. Vetiver grass cut at 4 week intervals is optimal in terms of output and nutritive value (Figure 3.3).

The study on the toxic content in the 10 vetiver cultivars reveals that the grasses have insignificant levels of nitrate and hence are not harmful to animals. Furthermore, nitrate and hydrocyanic acid elements are not found in vetiver.

Compost

Vetiver tillers and leaves are a good quality compost; in fact after 120 days the carbon/nitrogen ratio (C/N ratio) of vetiver decreases from 91–125 before decomposition to 38.9–47.5 after decomposition. The rate of decomposition gradually declines at 60–120 days. Vetiver tillers and stems are completely decomposed to become soft, disintegrated and dark brown to black in colour.

The analysis also showed that vetiver compost gains more major nutrients from the decomposition process, i.e. nitrogen, phosphorus, potassium, calcium and magnesium on an average of 0.86, 0.29, 0.12, 0.55 and 0.41% respectively, and has a pH 7.0 value. In addition to the above-mentioned major nutrients, vetiver compost also provides humic acid.

Biomass

Vetiver, owing to the fact that it is a type C_4 herbaceous plant, with great efficiency in terms of energy conversion by way of photosynthesis, has a high return per hectare of dry material, usually from 20 to 40 tons per year but there is an irrigated farm in Texas where it is claimed that the return is equal to 100 tons per hectare per year.

The calorific value of the biomass varies between 15 and 18 MJ/kg, equal to approximately 60% of the calorific value of carbon (from 24 to 33 MJ/kg) and equal to approximately 45% of that of fuel oil (40 MJ/kg).

Handicraft

Dry vetiver leaves are used for making handicraft items such as flowers (bouquets from vetiver leaves, flower baskets), ornaments (hats and belts), household decorative items (picture frames, mirror frames, fans, tissue paper boxes, and baskets) and toys and models (human dolls, animal dolls). The aromatic vetiver roots are used for making fans, clothes hangers and are mixed with other kinds of flower scents and leaves for making perfumed sachets (Figure 3.4).

Different uses of vetiver in the Mediterranean area

In the light of vetiver's characteristics which make it so desirable, the use of this species has been suggested for the Mediterranean basin, an extraordinary testimony to the value of the plant in the struggle against soil erosion and in the production of essential oil.

A project, lasting 30 months (from January 1994 to June 1996) and financed partly by the European Union, Directorate General VI Agriculture, Regulation 4256/88, art. 8, has been carried out in the Segura region (Murcia, Spain), one of the European regions most exposed to the phenomenon of soil erosion. This project was carried out on the "San Julian di La Hoya" farm, near to the city of Lorca. The farm is about 600 hectares in size; the main crops are almond trees in the hilly regions (the largest part of the farm) and cereals and oleaginose in the flatter areas. There are about 30,000 vetiver plants from Malaysia and US, previously grown in specially arranged nurseries, occupying an area of about 12 hectares.

The plants were employed against erosion in the following areas:

Figure 3.4 Handicraft products made with vetiver: hats, bags, household decorative items, flower bouquets, etc.

embankments (with a gradient of around 60%) (Figure 3.5);
terrace borders (Figure 3.6);
almond tree plantations (with a gradient of around 5%) (Figure 3.7);
gullies and canals;
and areas subjected to superficial microerosion.

Figure 3.5 Vetiver plants are employed against erosion in embankments with a gradient of around 60%.

Figure 3.6 Terrace borders with vetiver grass.

Plants were also cultivated for their roots in order to extract the essential oil.

In general, this pilot project has confirmed the survival rate of the plants; more than 89% of all transplants were successful even in not ideal climatic conditions. The root growth rate reached 1.7 metres in depth after 9 months and 2.6 metres 14 months after transplantation (Figure 3.8). The speed with which the tillers grew was such that after 9 months they had become, on average, about 60 in number originating from cuttings with around 5 tillers. The biomass yield was equal to approximately 40 tons/

Figure 3.7 Vetiver can also be planted along with other plants without disturbing growth or competing with nutrients.

hectare of dry material 14 months after transplantation. Vetiver's capability in soil protection was confirmed; for example when put to use in the protection of the escarpments it held approximately 20 cm of soil after only 12 months. Vetiver's sterility was investigated; the pilot project in progress in Murcia has confirmed that the plant does not produce any inflorescence after having been cultivated for three seasons. Its resistance to winter climatic conditions was studied. Vetiver showed no signs of frost injury even at temperatures of −2 °C. Its oil producing potential was evaluated; in roots 12 and 16 months old a yield of approximately 1–2 per cent was obtained, with good quality certified by the organization responsible for certification and quality control.

The project has furthermore shown that vetiver, despite the fact that it originates in tropical and sub-tropical areas, adapts well to temperate climates. Vetiver can therefore be used to advantage in Mediterranean areas and in the numerous marginal areas

Figure 3.8 Vetiver root growth rate can reached 1.7 metres in depth after 9 months and 2.6 metres 14 months after transplantation.

of the European Union, which have been abandoned by classical farming and condemned to waste, leading either to a situation of progressive desertification or to the uncontrolled proliferation of weeds.

The results obtained can be transferred to other Mediterranean regions, allowing marginal areas to be protected without imposing on the income of the farmer, or on public funds; indeed the extraction of essential oils from roots, the exploitation of its biomass and manufactured products more than compensate for the costs of the plantations and the fight against erosion, without quantifying the agricultural, environmental and social benefits of this agricultural technique.

Planting Techniques

Vetiver can be used both for soil conservation and for the production of essential oil.

The two techniques present important technical and economic synergies which are described in the following sections of this chapter.

Planting for soil conservation

The planting of vetiver can be successfully achieved in numerous situations, performing its function in line with its various morphological and pedoclimatic characteristics, owing to the fact that the plant can adapt outstandingly well to extreme situations.

The vetiver cultivated in Murcia, starting from imported mother plants, is completely adapted to the local soil and climate conditions and shows excellent characteristics as an environmental tool against soil erosion. Its root system grows quickly in depth and with features not present in other plants. The well developed root mass can attain, in certain cases, 5 m in depth. From the roots it is possible to extract, a couple of years after planting, an essential oil with good organoleptic characteristics demonstrated by analyses and official tests.

An outline of the most significant types of measure is presented in the following paragraphs.

Localized

This refers to localized and specific measures. In order for this to be achieved it is not necessary to have many vetiver plants, indeed often only a dozen or even single plants are used. This concern measures on the scale of using individual trees, for example for soil conservation in an orchard, or for conservation of certain sectors of old terraces built with dry stone. It is possible to take action to consolidate the narrowest and steepest passages which follow the line of the valley, and which collect rainwater. The use of vetiver also efficiently protects bridges and run-off water channels.

Linear

This refers to the linear planting of vetiver used for a specific type of protection. For example it is used in the protection of fields in any location or along terraces, right above the retaining wall. The continuous vetiver hedge is also used along canals, streams or rivers in a particular way near to the bends, where the erosive force of the water is at its greatest. In the same way vetiver is planted to protect paths or roads by planting the hedge both upstream and downstream. Furthermore it can be used for the conservation of shores of lakelets, artificial embankments and little canals for irrigation or water drainage. The vetiver hedge can also efficiently slow the flow of the river, when planted across the river itself. This type of measure is particularly well suited to rivers in Mediterranean areas.

Figure 3.9 The distance of 1 vertical metre between the rows (height) is recommended in literature and has produced outstanding results in erosion control in the project carried out in Murcia.

Extensive

When the surface area on which the action is being taken is large and erosion problems affect the whole area, it is necessary to plant the vetiver hedges along the contour lines for soil conservation. Often it is a good idea to plant near to natural or artificial reliefs, for example near to terraces, trees, boulders or outcropping rocks.

Plantation density

Plantation density depends on the type of soil conservation (localized, linear or extensive) and, obviously, on the type of site which requires these soil stabilization measures. The vetiver hedges are planted orthogonally to the direction in which the water on the soil is moving.

In order to ensure that the measure to contain the erosion is effective, vetiver must form a tight hedge: the further the plants are away from each other, the longer it will take for the hedges to close. For example, if the hedge is required to close in 4–5 months, the plants must be positioned approximately 10 cm from each other. If one can wait 10–12 months they can be planted at approximately 20 cm intervals.

For the distances between the rows the fundamental factor is the slope of the soil. As the slope increases, the distance between the hedges must decrease. In the pilot project the vertical distance (height) between the rows was 1 metre with a linear density of 10 plants per metre. The distance of 1 vertical metre between the rows (height) is even recommended in literature and has produced outstanding results in erosion control in the project carried out in Murcia (Figure 3.9).

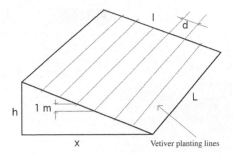

Figure 3.10 Definition of the various sizes.

In order to facilitate the use of this table remember that the various sizes (gradient of the soil, width, length, etc.) are defined as according to Figure 3.10, where **p** = gradient % = 100 × **h** / **x** (a gradient of 100% corresponds to an angle of inclination of 45°); **x** = horizontal length; **h** = height; **d** = distance between the rows; **l** = soil length and **L** = soil width.

The formula for the geometric calculation of the distance between one row and the next is:

$$d = \sqrt{1 + \frac{10000}{p^2}}$$

As time goes by the closed vetiver hedges make an ever increasing terrace, due to the collecting of the eroded soil on the uphill side. These terraces are the ideal habitat for spontaneous species which further stabilize the area. Furthermore the vetiver hedges slow the water run-off, reducing erosion and allowing the water to soak into the soil, improving soil moisture and therefore increasing the local water resources.

Preparation of the soil

The planting techniques depend on the prevailing conditions, such as the slope, the nature of the soil, the distance from roads or water sources, etc.

In a situation where erosion exists and there is serious water deficiency, mechanical intervention can be problematic and therefore it is often necessary to continue manually. The planting site is prepared by manually digging trenches of the correct dimension for the vetiver cuttings (width 10 cm, depth 20 cm) (Figure 3.11). In the case of the slope not being excessive and the stability being fairly good, the use of mechanical equipment, such as ploughs, diggers or even simply motorized hoes is possible (which facilitate subsequent work).

If necessary, according to a soil analysis, the spreading of fertilizer would correct any severe shortage of minerals. Such a measure would be undertaken at the same time as transplantation. Subsequent fertilizations are not necessary, unless the conditions specifically require it.

Figure 3.11 Tranches of the correct dimension for vetiver cuttings.

Planting and management

The transplantation is achieved manually, planting the cuttings, prepared in poly-styrene bags, at regular intervals of 10–20 cm apart.

After transplanting it is prudent to provide for emergency irrigation. It is further prudent to ensure that transplanting takes place during the optimum season, with well developed plant material so as to ensure a high number of cuttings successfully taking root.

In the pilot project carried out in Murcia, the optimum period for plantation turned out to be in March, when there is no longer the danger of frost. Likewise the autumn period, before the winter temperatures set in, has turned out to be a good time to plant (September to mid-October) so that the plant continues to grow and, at the same time, continues to put out its roots, as long as there is sufficient moisture in the soil.

Weed control is not usually necessary as soil with a severe shortage of water and erosion problems does not make fertile ground for local vegetation. Once vetiver has become well established, the presence of other species of plant allows for an improvement in soil stability: the other herbaceous plants do not present significant competition for vetiver which has a robust nature and a deep root system (even more than 5 m).

In the first cultivation periods it is important to sort out any crop failures and provide the necessary water supply in particularly arid areas: owing to its deep root system vetiver is able to provide itself with water from the moisture deeper down in the soil within the space of about a year.

Other specific measures are not necessary. Only as the winter months draw nearer is it necessary to cut the leaves off, so that the plant can face up to the challenges of winter in the best way possible and facilitate its subsequent recovery.

Planting for oil extraction

The production of the essential oil implies the cultivation of vetiver for at least two years, with a view to obtaining a sufficient yield of root biomass with optimum characteristics and then to proceed to the distillation process.

Preparing the soil

Planting is carried out in plots on level ground, which are medium textured, deep and without skeleton, rich in organic matters and nutrients and irrigated if possible. In these conditions the roots can develop in depth without encountering obstacles. The soil must be prepared by deep ploughing (30–50 cm) and crossed tilling, which refine the structure of the soil.

Planting density

In general, a density of approximately 40,000 plants/ha is considered optimal and is obtained by planting at intervals of 80 cm between the rows while the cuttings in the rows are planted at 30 cm intervals.

Planting

Vetiver culms with short roots (3–5 cm) and with approximately 20 cm of leaves are used. The transplantation is carried out manually or with mechanical equipment which make the task easier, consisting of tractors which carry the workers on a special frame. Planting is generally performed in the spring when air temperature is constant at about 10 °C. Before planting, it is wise to use fertilizers on the soil, so as to ensure its adequate fertility, depending on soil analysis and vetiver requirements.

On the basis of the experience in Murcia, it has been determined that 75 units per hectare of base nutrients (N, P_2O_5, K_2O) using a 15.15.15 fertilizer are sufficient for a positive effect. Weed control is important in the first phases of plant growth. This is easily accomplished by using a herbicide, like simazine, before planting vetiver and even before weed emergence.

Cultivation

Suitable fertilization measures are scheduled depending on the development of the plant. They, in general, consist only of 300–500 kg/ha of potassium nitrate applied several times (at least 4) in the course of the growth period (April to October). The same operation is carried out in the second plant cycle. When planting is done for root production and subsequent extraction of the oil, the water supplies allow optimum growth of large quantities of roots in short periods of time with optimal oil production.

From the irrigation methods available, either sprinkle or trickle irrigation can be used, according to the particular needs.

During the course of the project in Murcia, with rainfall at a rate of 250–300 mm and potential evotranspiration of 9,500 m^3/ha (according to Thornthwaite), excellent results were obtained with an annual irrigation equivalent to 6,000–7,000 m^3/ha.

Weed control is only important in the planting phase, since vetiver then closes together completely between the rows, stifling the growth of infesting elements. After having carefully prepared the soil for transplanting it is possible to scatter an anti-germinating product so as to kill any weeds that are present in seed form.

After weed emergence, it will be necessary to use specific products. It is also possible to carry out superficial tillage between the rows. For this reason it is wise to engineer the distance between the rows to match that required to perform this task for example for Indian corn and sunflower (see Figure 3.9).

During the course of the season a few vegetation cutting measures are undertaken, designed to contain growth. Approximately 20 cm of stem is left behind, in order to make subsequent work easier. Because vetiver leafy lamina contains high levels of silica, the cutting equipment suffers high levels of wear. For this reason it is necessary to sharpen the blades often or use long-wearing blades. Cutting vetiver means that the residue, such as the mulch, can be redistributed between the rows. The cut vegetation can also be efficiently used as biomass for various uses, in particular in generating energy. As has previously been mentioned, other measures in particular are not required. Only as the winter months approach do the leaves need cutting back, so as to allow the plant to face up to the challenges of winter in the optimum way and to facilitate re-growth in the spring.

Root harvesting

This can be done in any season, both when there is a need to produce the oil, or when the plant material is being prepared for soil conservation. Naturally, the earlier the

Figure 3.12 Pulling out of the soil of vetiver plants.

plant is dug up, the lower the quality of the roots. For this reason it needs to be pointed out that the roots must be dug up at the end of the summer, starting from the second year of cultivation when the roots will have grown considerably and the air temperature will still be sufficiently high to dry the roots effectively. The soil must not be too moist in order to avoid collecting excessive soil residues with the roots. First the vegetation is cut and the cut material is collected to allow the machinery to continue with the subsequent phases. Approximately 20 cm of stem is left, in order to make the subsequent phases easier. Next the cuttings are dug up. This operation can be achieved with machines like tuber or root diggers. The digging is done to a depth of approximately 60 cm, where the majority of vetiver roots are concentrated. The roots, once having been undermined and pulled out of the soil, are separated from the crown, using shears or circular saws (Figure 3.12).

On average, root production is 2–4 t/ha. Assuming an oil yield of 2–4%, 40–80 kg/ha of essential vetiver oil is produced.

The production plantation, if not a nursery in the middle of a field, allows optimum material to be produced. The collected culms (40–60 culms per plant) can be separated from the others and used for subsequent transplants. The root production plantation can therefore also act as a nursery for the reproduction of vetiver, with clear savings in cost and labour.

Reproduction Technique

Vetiver was introduced into Europe by the company "Tecnagrind S.L." from whom the plants are obtainable.

Reproduction of vetiver by separating the clumps

In view of the vetiver high reproduction potential, it can be worthwhile organizing the reproduction of planting material on the farm. The culms which have sprouted can easily be separated by hand, opening out and stripping the leaves off the clump and subsequently transplanting into the field.

Rooting in a pot

When planting for soil conservation, the use of good quality material, well rooted in polystyrene bags or possibly in pots, allows for a higher percentage of plants to take root, greater speed of conservation and less failures. The material can be transplanted into the field when there is no longer a risk of frost. The most suitable containers are of hard plastic, square, 7–8 cm in width and 18–20 cm deep, with a capacity in the region of 1–1.5 litres. The ideal potting material for growth consists of a mixture of peat, sand and soil with a pH close to 7. This is mixed with a substrate of 1.5–2.0 kg/m^3 of long release fertilizer (6–8 months).

Farming dedicated to the plant(s) grown in pots is similar to that dedicated to any other green species. Vetiver cultivation requires frequent irrigation to maintain an adequate level of moisture in the substrate and nitrogenous fertilization for 2–3 weeks during the growth cycle. Pruning to a height of 30–40 cm encourages growth of the plants and strengthens the clumps. Due to the lack of pests or diseases in Mediterranean areas, no phytosanitary measures need be taken.

Storing

The place built for root and leaf storage must be cool, well ventilated and clean, in order to avoid the development of moulds and bacteria or contamination related to mouse or other animal defecation. This is important to maintain a high quality level, to avoid loss of a part of the harvest and to warrant the conformity to the standard sanitary rules of the raw, semi-manufactured and manufactured products. The store is composed of an uncovered area of about 300 m^2 where it is possible to lay the roots after extraction from the soil and to allow them to dry naturally until the optimal humidity degree (15%) is reached. This is about 10 days in the climate present in the Murcia region in November. The store has also a covered area of about 150 m^2, that had been employed as a wine vault for wine aging at the farm where the pilot project took place. This choice has been made in order to get a cool and dry place where it is possible to store the roots before distillation.

The store also possesses electrical power, running water at environmental pressure, and enough ventilation for correct root and final product storage. These facilities are also useful for perfect operation of the mobile distillation unit.

Oil Extraction

General comments

Vetiver oil is currently produced in tropical countries (Giava, Réunion, and Haiti) by local perfume industries who make healthy profits exporting the oil; the roots are also exported and are then processed in destination countries, including European countries. In this way the added value derived from the production and sale of the essential oil does not remain with the farmer but is transferred to the perfume industry.

During the course of the project in Murcia, with a view to increasing European farmers' incomes, a portable oil extraction system was set up, capable of carrying out oil distillation on the farm itself. Furthermore, during the course of the project simple handicraft products were made and sold directly on the farm with considerable economic advantages for the farmer.

As far as distillation is concerned it can be noted that it is possible to make vetiver oil production systems with different production capacities in order to satisfy the most diversified of the farmers' needs. The system that was set up during the course of the project satisfied the requirements of approximately 5 farms of approximately 5 hectares each. The mobility of the distillation equipment allows distillation to be carried out directly on the farm with important advantages and savings in cost.

In order to further decrease the costs of the production process which is already relatively low given the fact that the cost of mass-produced distillation equipment could be around 600–1800 ECU according to production capacity, the distillation equipment can be purchased by a small co-operative, loaded on to a small vehicle and transported to where it is required by the members of the co-operative.

In order to allow the farmers periodically to analyse their oil, a small laboratory is also available to check, measure and analyse the quality of the vetiver oil. The instruments, identified after careful market analysis and after numerous equipment tests, are small, low in cost and can be placed in a small space in the co-operative storage building. The analytical equipment requires little electrical power (maximum consumption is equal to that of a 50 watt bulb) and is user-friendly.

A good distillation method was developed as the result of a huge number of trials, tests, measures, accomplishments, researches and simulation directed to the creation of a distillation system with the following characteristics:

- good distillation efficiency;
- high safety;
- small dimension;
- easy transport;
- high resistance;
- easy maintenance;
- low mass-produced price;
- low energy consumption;
- possibility of operation without employing external power, eventually utilizing the heat generated by combustion of the vetiver aerial portion;
- low water use;
- very low environmental impact;
- analogy with known techniques used in the countries of origin;
- oil quality;
- necessity of qualified personnel employed for the operation;
- easy availability.

Extraction technique

On the basis of the above comments there now follows an explanation of the extraction method which was set up in order to make it easier for the farmer or the co-operative wanting to carry out the oil extraction, either for themselves or for a third party. The technique is easy to master and can be adapted to the most diverse production needs.

The operations which must be carried out are summarized as follows:

- trituration of the roots;
- distillation of the oil;
- decantation and dehydration;
- testing.

Trituration of the roots

The preparation of the roots is a simple operation. Once the roots have been de-earthed and washed, they are dried preferably in the shade. The drying period is not critical even if it is advisable to wait a few weeks so as to obtain an oil which has a warmer aroma as the most volatile compounds which are lost in part during the maturing of the roots in the air, give the essential oil a drier tone. The roots, cleaned and dried, are easy to cut. The roots must be cut up into pieces, at the most a few centimetres in length. It is possible to employ the normal shredders used for wood or plant material (Figure 3.13).

It is a good idea not to allow the roots to become excessively overheated during the cutting operation and it is therefore advisable to use a blade, as opposed to a hammer

Figure 3.13 It is possible to use normal shredders for wood or plant material to cut the roots in small pieces before distillation.

shredder. Even a normal, low-cost, garden shredder (approximately 300 ECU or less) can easily be used. With a machine such as this it is possible to produce approximately 10 kg/hour of roots with a single operator and the machine consumes approximately 1.5 kW of electricity.

Distillation of the oil

A careful choice of manufacturing methods giving the best technical results has been provided, in order to demonstrate that the important and vital function of environmental protection of desert lands could finance itself through the simple and cheap transformation of the plants employed. Seven distillation systems have been analysed accurately and a merit list strictly related to a particular employment, which requires a simple and cheap system, has been drafted. This system must be easily manageable and maintainable and must require low quantities of energy and water.

Many distillation processes are available:

- distillation with water and steam at environmental pressure using direct fire;
- direct fire distiller surrounded by a jacket filled with insulating oil and cohobation of the distillation water;
- distillation employing only steam at environmental pressure;
- same procedure analysed above, but using 1.5 bar pressure;
- turbodistillation;

- solvent extraction;
- supercritical CO_2 extraction.

A comparison of all the methods employed follows.

The direct fire distiller employing water and steam at environmental pressure is composed of one alembic filled with water and plant material, warmed up by a flame in direct contact with the lower wall of the boiler. The distillate is cooled down by using a water cooling coil and collected into a container. The oil floats on the surface or deposits on the bottom and it is collected through proper separating funnels or florentine flasks. This method is simple, economic and easy. However, a disadvantage is the length (36–48 hours) necessary for processing 300–400 Kg of plant material (Reunion). Also, this method can give a cooked smell to the oil, if the plant material overheats.

Based on experiments conducted by a Haitian furnisher of vetiver using the method of water and steam distillation at environmental pressure (20 hours for processing 400 Kg of roots meaning 20 Kg/hour), it is known that the percentage of oil in comparison with the total distillate is 0.2%.

This percentage is much higher than that obtained from repeated experiments by von Rechenberg whose results were published in 1910. In that case, the percentage was 0.015–0.020%. The difference is related to improvements in the apparatus construction techniques. The oil yield represents 2.5% in weight with respect to the roots. This yield is very high and depends on the excellent quality of the plant material employed.

A small portable distillation plant, constructed of copper (see also comments on copper later in the chapter) and tin and direct fire on a 70 litres boiler, could cost, if mass-produced, about 800 ECU. Employing a similar plant, every time, it is possible to distil about 10 Kg of root material in 40 litres of water. It is necessary to open the plant in order to add water during the distillation process. The water cooling consumption is about 6 l/min.

The distiller surrounded by a jacket filled with insulating oil and cohobation of distillation water represents a better version of the distiller previously described. In particular, the jacket filled with insulating oil assures a uniform distribution of the heat into the boiler and protects plant material from overheating.

Besides, the cohobation (the re-infusion of the more aqueous fraction of the distillate into the boiler through a suitable duct) avoids the necessity of water addition into the boiler every 2–3 hours, as is necessary using the method described above.

This procedure allows increase of the distillation efficiency and the percentage of oil with respect to distillate. This kind of instrument is easy to use and more economical than the direct fire distiller, and it is usually the favourite in the artisan extraction of essential oils from herbs. A system similar to that described above, at least in dimension, could cost, if mass-produced, about 900 ECU. It is possible to recover the cooling water and there are no pollution risks.

The steam distiller employs a steam flux insufflated into the boiler, thus allowing the extraction to proceed. Steam is produced by an auxiliary steam generator. This method possesses a greater productivity, but it requires a generator of white steam for good use, which must be always kept in first rate working condition, therefore employing technical personnel. Hence, its management is more complicated in comparison with the two equipments previously described. In order to employ the mobile unit in several areas lacking electrical power, it is necessary to add the cost of a movable diesel electricity generator (20–30 kW).

The distiller at pressure higher than environmental pressure is an instrument using dry steam flux at about 1.5 bar. This kind of equipment shows some safety problems related to the joints and closure flanges. In this machine plant material is maintained at more than 100 °C, and it is necessary to check the temperature in order to avoid exceeding the temperature at which essential oil degradation starts. For vetiver, in particular, one must not exceed 120 °C.

Experiments performed using a distiller working at 1.5 bar and 120 °C on vetiver from Haiti have shown a reduction of the length of operation to 8 hours for processing 400 Kg of roots (50 Kg/h). The yield percentage on distillation was 0.4% of oil.

In order to employ the mobile unit in several areas lacking electrical power, it is necessary to add the cost of a movable diesel electricity generator (20–30 kW). Also, the cooling water consumption increases in proportion to the root quantity. For this reason 100 l/m of water are necessary for processing 400 Kg of plant material.

The turbodistiller is a water and steam distiller working at environmental pressure, but it is modified by the addition of a rotating blade inside the boiler that, operating at high speed (500 rpm), keeps on cutting the plant material in the water and spreads uniformly the steam flux within the boiling mass. This method, combining the action of cutting and mixing, makes the process faster. However, it requires equipment of a higher technical level, owing to the presence of a driving shaft within the boiling water. Furthermore, it needs the employment of special ball-bearings and seal-valves that can present great strains related to the cutting and vibrating processes.

Experiments done using a turbodistiller working at environmental pressure on Haitian vetiver have shown reduction in the length of operation to 5 hours for processing 400 Kg of plant material (80 Kg/h), with a resultant percentage of 0.5% of oil on completion of distillation.

A pilot portable plant for turbodistillation, able to treat 80 Kg of plant material per hour, could cost more than 350,000 ECU. In order to employ the mobile unit in several areas lacking electrical power, it is again necessary to add the cost of a movable diesel electricity generator (20–30 kW). The cooling water consumption is very high (about 4 mc/hour) and this system requires specialized technical personnel.

The solvent extractor is based on the principle that essential oils are soluble in organic or inorganic solvents. The oil must be separated from the solvent (i.e. petroleum-ether) with subsequent distillation processes up to complete purification. To speed the process up, distillation is made at reduced pressure. A typical pilot plant working with 2–3 Kg/hour of raw minced material can cost about 230,000 ECU (portable version).

In order to employ the mobile unit in several areas lacking electrical power, it is necessary to add the cost of a movable diesel electricity generator (20–30 kW). Owing to the complexity of the system, a specialized technician is also required. According to law, periodical overhauls and checks are necessary in order to avoid accidents related to the parts under pressure. Explosion risks, depending on the solvent employed, are also possible.

The extractor working with supercritical CO_2 is a solvent extractor that uses supercritical CO_2 as a solvent. This is a method developed because it ensures a better quality of oil, and in the meanwhile avoids contamination problems of product and environment. It is employed in the extraction of caffeine from coffee. It is possible to

Table 3.2 Features adopted for the choice of the optimal distillation system to employ in agriculture.

CHARACTERISTICS	DISTILLATION SYSTEMS ANALYSED						
	D1	D2	D3	D4	D5	D6	D7
• good distillation efficiency		✪	✪	✪	✪		
• high safety	✪	✪					
• reasonable dimension	✪	✪	✪	✪			
• easy transportability	✪	✪					
• high resistance	✪	✪					
• easy maintenance	✪	✪					
• low mass-production cost	✪	✪	✪	✪			
• reduced energetic consumption	✪	✪					
• possibility of operation in absence of electrical power, using the heat produced by combustion of Vetiver aerial portion as a feeding source	✪	✪					
• reduced employment of water		✪			✪	✪	✪
• reduced environmental impact	✪	✪	✪	✪	✪		✪
• analogies with known techniques employed in the countries of origin	✪	✪					
• quality of the oil produced	suff.	good	good	good	good	excel	excel
• specialized personnel not required	✪	✪					
• easy availability	✪	✪	✪				
• FINAL JUDGEMENT (E = excellent; G = good, S = sufficient; I = insufficient)	E	O	S	I	I	I	I

D1 distillation using water and steam at environmental pressure and direct heating;
D2 distillation using water and steam at environmental pressure and direct heating surrounded by a jacket filled with insulating oil and cohobation of the distillation water;
D3 distillation using only steam at environmental pressure;
D4 identical to D3 but using a pressure of 1.5 bar;
D5 turbodistillation;
D6 solvent extraction;
D7 distillation using supercritical CO_2.

realize this machine as a mobile unit able to treat about 3–3.3 Kg/hour of plant material. Its cost is about 300,000 ECU but in order to employ the mobile unit in several areas lacking electrical power, it is again necessary to add the cost of a movable diesel electricity generator (20–30 kW).

Legal restrictions on employment and maintenance are valid and the same considerations as reported above apply.

Table 3.2 describes the rule adopted for the choice of the optimal distillation system for employment in agriculture.

The best distillation system is the D2, the water and steam distiller operating at environmental pressure by direct fire heating surrounded by a jacket filled with insulating oil and by cohobation of the distillation water. The turbodistillation process requires high financial support, shows a great productivity with regard to the preset employment, and it is suitable for use with 400–500 Kg of dried roots or more. This system needs periodical and expensive checks for the steam generator and seal-valve maintenance. For this reason highly specialized personnel are required. This method is suitable for the construction of large extraction stations.

The use of a simple water- and vapour-alembic, which is not under pressure, allows optimum distillation for the production of small and medium quantities of essential oil without requiring large amounts of heat energy and above all without using large quantities of cooling water. The size of the alembic depends on the quantity of roots which are required for processing at the same time.

Alembics with a total volume of between 50 to 150 litres are easily manageable both with modest energy and water resources and are low in cost. Considering for example that a 70 litre alembic costs approximately 900 ECU (when mass-produced), it allows around 10 Kg of roots to be processed at a time thus producing approximately 0.1–0.2 kg of oil, using a simple 10 kW gas stove and a supply of cooling water easily obtainable from a normal household tap (6–10 litres/minute at 15–20 °C, possibly recyclable in closed circuit). This alembic consists of the main stainless steel chamber as vetiver oil is not compatible with copper. The chamber is surrounded by a jacket filled with diathermic oil (so in order to homogenize the internal temperature and ensure that the roots do not overheat locally). The condenser and the condenser's collection burette complete the alembic unit.

Distillation process

20–30 litres of tap water are poured into the chamber. Then a steel grille is put into place, on which the cut roots are placed. On top of this, a second grille is placed in order to hold the roots in place. The roots during the process tend to swell and could block the vapour from escaping from the lid of the chamber via the escape duct. Once the lid of the chamber is closed and the stove has been lit the temperature must reach approximately 103–104 °C before distillation can begin. The roots are partially wet from the water and vapour passing through and extracting the essential oil. The vapour which has been conveyed to the condenser condenses in the collection burette. Here the essential oil is collected, as well as the condensation water, which is then reintroduced into the chamber via a recirculation pipe. The vetiver is amber-yellow in colour with various shades from green to red according to the type of roots and their maturity.

There are two fractions, one lighter than water and highly aromatic, which is collected in the upper part of the burette, and one which is denser than water, is wax-like and clear, and collects in the bottom of the burette. Both fractions are important for the organoleptic composition and therefore for the quality of the oil as a whole. The process lasts approximately 12–15 hours and the burette must be emptied every 2–3 hours.

The total liquid gas consumption for one extraction is approximately 5 kg. During the process only minimal supervision by a worker is required to check that the temperature does not get too high, that the gas and the cooling water behave normally and finally, to collect the condensed material at certain intervals. The collected liquid is composed of the two vetiver oil fractions and water.

Decantation and dehydration

The extracted liquid is left to decant for approximately 12 hours in a glass container, and therefore a pipette is used to collect the two oil fractions (from the base of the container and from the surface of the liquid). The oil collected in this way still contains traces of water which are inevitably also collected by the pipette.

If required, the oil can be separated from the residual water by adding anhydrous sodium sulphate (8 g per 100 g of oil) and then by filtering it all through filter paper. By this process, which lasts a few hours, perfectly clear, water-free oil can be obtained. The oil produced should be stored in dark glass containers to protect it from the light.

Testing

The appraisal of the quality of the oil produced allows the farmer to correlate agricultural practices (type of plant, type of cultivation site, the season and the age of the plants) with the quality of the oil produced, and therefore to gain the necessary experience with which to optimize the product itself. The commercial qualification analysis of vetiver oil has been reproduced in Table 3.3 along with important characteristic values.

Commercial rules and **quality standards** related to vetiver oil parameters have been developed, in order to make a comparison between requested values and values obtained from experimental cultivation. It is necessary to have an equipment suitable for making the most important analyses of oil quality and international experts able to carry out tests and certify controls. The main characteristics of the best oil in commerce, which means the oil extracted in specialized distilleries starting from dried roots and imported from the countries of production, are described in Table 3.3.

The data reported in Table 3.4 was obtained with analyses carried out on oil extracted using the best distillation system and 18 month-old roots coming from the experimental cultivation in Murcia.

Table 3.3 Main characteristics of the best vetiver oil in commerce, which means the oil extracted in specialized distilleries.

– rotatory power	from +15° to +45° (from +13° to +17° for young roots) at 20 °C
– refractive index	from 1.52 to 1.528 (from 1.515 to 1.527 for young roots) at 20 °C
– carbonylic compounds	with PM 218 from 9 to 26.5%
– acid index	from 27 to 65
– ester number	from 9.8 to 23
– esterification number	After acetylation 110–165
– free alcohols	with PM 220 from 21 to 60%
– gas-liquid chromatography	with chromatogram corresponding to that reported in literature
– density	comprised between 1.015 and 1.040 (between 0.990 and 1.020 for young roots) at 15 °C; between 0,982 and 0,998 at 30 °C
– solubility in 1–3 volumes of 80° alcohol	solution becomes cloudy if more alcohol is added
– colour	from yellow to yellow brown (according to several authors the oil can assumes an amber-yellow colour, dark blond, reddish brown, dark brown, almost black); a very dark colour is related to the presence of heavy metals or metal impurities
– odour	Full-bodied scent, rich, sweet, woody and very persistent (green, earthy or sour if the roots are young), improves following aging. Regarding the olfactory characteristics aging is believed mandatory in the opinion of recognized experts in the perfumery field

Table 3.4 Certified analyses on oil extracted from 18 month-old roots.

Type of analysis	American vetiver 18-month old	Malaysian vetiver 18-month old
– rotatory power	+36.80	+40.00
– refractive index at 20 °C	1.5217	1.5214
– carbonylic compound % (such as vetiverone)	16.55	11.99
– acid index	6.77	5.75
– ester index	19.34	23.23
– chromatographic analysis	peculiar	peculiar
– vetiverol %	9.64	9.93
– nootkatone %	0.5	1.27

Attention must be drawn to the fact that such analysis is generally aimed at exposing possible commercial adulterations, and is only possible with sensitive and expensive equipment used by trained personnel. However some of these can be conducted with apparatus that is low in cost and easy to use, and they can give significant results, especially if they are correlated with the organoleptic analysis and with the colour. The significant areas to be analysed are organoleptic analysis, refractive index, solubility in alcohol at 80° and density. Organoleptic analysis is undoubtedly the most direct and accurate way in which to estimate quality. This is performed using thin strips of absorbent paper, known as "touches". The analysis is of a comparative nature. A "touche" is lightly soaked in a vetiver oil of known quality (a sample of Réunion oil or Haiti oil from the previous collection) and a second "touche" is soaked with the oil to be tested. By smelling first one and then the other in succession, different olfactory tones can be distinguished and therefore quality assessed. For this type of analysis it is clearly necessary to have a trained "nose"; it is, however, very easy for the farmer to gain experience in this area. The analysis is clearly of a subjective nature and an average can be found by other people. The measurement of the refractive index is achieved with the help of a simple refractometer costing approximately 300 ECU. One or two drops of the oil to be analysed are smeared on the instrument's glass slide; looking through the eyepiece against the light, the refraction line on the internal scale is to be seen by a colour change. The measurement must be correlated to the temperature of the environment by means of a correction table supplied with the instrument and easily read. The solubility in alcohol at 80° is achieved by using a small calibrated glass in which one part of oil and one part of alcohol are mixed and the transparency of the solution obtained is observed. By adding subsequent parts of alcohol, the solution becomes less transparent. For a good quality oil this occurs when the fourth part of alcohol has been added.

Density measurement is the most complex method, owing to the nature of vetiver oil which has a density very similar to that of water. Measurement can be carried out by using a small calibrated glass or pycnometer, with water which has been distilled twice, and with analytical scales with a resolution of 0.001 g. First the water (which has been distilled twice) is weighed and then the same volume of oil; thus the density value is obtained by dividing the two measures. This value must similarly be corrected as regards the temperature of the environment by means of a suitable correction table.

Storage

The essential oil must be stored in dark glass bottles in a cool and shady place. The bottle cap can be made of Teflon® or polyethylene. It is possible to use iron containers, coated with zinc, but not those copper-coated. As the oil is a good rubber solvent, rubber caps must be avoided.

Uses of vetiver oil

The farmer can, in a very simple and economical way, make many products with vetiver oil, working within his own farm as well as earning a respectable wage through sales.

As outlined above, these products can generate a reasonable commercial enterprise in conjunction with farm holidays and with promotional or tourist activities at local or regional level. Using the roots directly it is possible with a few grams of the roots to make small fabric bags used as deodorants for small enclosed areas (cupboards, drawers, cars, etc.). Vetiver has a strong and pleasant aroma and furthermore, the presence of insect-repelling substances in the oil has been scientifically proven. With only a few drops of essential oil it is possible to make the following products:

- household deodorants (for example little bags of dried flower petals scented with a few drops of oil, or small blocks of clay also soaked with only a few drops, or scented, natural wax candles which clear the air of smoke);
- personal hygiene products such as aromatic soaps or shampoos, similarly made with products with a neutral aromatic basis and with only a few drops of oil added;
- household cleaning products; for example products used for cleaning the floors or the toilets, again using neutral commercial products and adding a few drops of essential oil;
- spray deodorants; by diluting the oil with water and alcohol, a first-class spray deodorant can be made using containers with gasless pumps. The spray can be used in deodorizing shoes, bags, suitcases, store-rooms and so on.

Soaps (Figure 3.14), shampoos, candles and scented solutions have been supplied to customers of farm holidays, herbalists shops, perfumeries, chemists shops, etc., after asking an opinion. In general, people had a positive impression, which means that they like the perfume, as the odour related to the basic compounds, the tallow odour in the basic soap, for example disappears and the vetiver note increases. There are some

Figure 3.14 Soaps prepared using vetiver oil.

divergent opinions on the pure odour of vetiver, anyway. Some customers like it, others think that it is too medicinal and they would like a scented mingled mixture in which the characteristic note of vetiver is conserved. In fact, perfumers are moving in this direction, producing particular mixtures (Vetyveral, for example) which are to be considered also natural identical compositions in which many scented essences are present. These mixture are able to increase the typical vetiver note. Farmers can repeat this experience, making their own compositions. The choice regarding the alembic type has been made in order to employ it also for distillation of other natural essences deriving from spontaneous or cultivated plants producing essential oils. Extraction procedures are similar to those previously described. This allows an increase in the number of essences available in the farm and to enlarge the offer of products.

Handicraft items produced with bleached roots and leaves have been given to farm holidays customers and distributed to the local population. Everybody liked them. The mat was the most interesting object, as it can substitute for a bathroom carpet, or a table plate and in the meanwhile is able to release a pleasant odour in the environment. The objects deriving from bleached leaves are identical to those manufactured with straw.

Vetiver in perfumery

Like a large number of exotic essences such as ylang ylang, rose wood, etc., vetiver is an essence used in recent years. It is necessary to wait until the end of the 19th century before seeing the name vetiver in the general catalogues of perfumery without mentioning its origin. Its particular odour was described at that time as warm and deep and similar to myrrh and to iris roots. It was employed as "essence", an essential oil that women bought in perfumery shops and mixed with other essences in a careful way, in order to create their own perfumes.

Although it was mentioned in the ancient formula of the "Bouquet du Roi", few perfumers employed the essential oil of vetiver at the beginning of the 20th century. Although it existed as a tincture or as a resinoid, factories manufacturing organic chemical products such as De Laire, Naeff, Givaudan, specialists in perfume chemistry, sold it as a vetiver acetate or vetiveryl acetate. Vetiver acetate is obtained starting either from the essence of vetiver or from vetyverol, the main component of the essential oil.

At the beginning of the 20th century, when the reference is to the product of natural origin, the perfume chemistry manufacturers created characteristic notes, which were considered difficult to use with their violent, hard, lack of gradation features. Several perfumers created bases, mixtures composed of about ten constituents, in which a characteristic note is rigged. In 1910, Marius Reboul, perfume composer for Givaundan, created a base of vetiver acetate, coumarin and bergamot named Sophora which was used in the composition of many perfumes, starting with Air Embaumé of Parfums Rigaud (1912). This is a flowered perfume, with pink, vanilla and vetiver essences which will become an archetype of modern perfumery.

Conclusions

Vetiver grass is a coarse perennial plant which adapts well to the Mediterranean climate, achieves a high percentage of plants that take root successfully, spreads rapidly and is not a potential weed. The roots develop rapidly in any type of soil, reaching a depth of over one metre in just a few months, and several metres after only a few years, thus making a natural underground barrier, effective in erosion control. Furthermore, a highly valued essential oil can be extracted from the roots.

The economic advantages which derive from vetiver can be subdivided into indirect advantages and direct advantages.

Indirect advantages

These refer to the economic, environmental and social advantages which derive from using vetiver. In fact, a plant of this type allows erosion control to take place without environmental impact and without the use of complex and expensive man-made constructions. These constructions are often damaging to the environment as rigid control of river and stream waters can provoke dangerous and subsequently disastrous floods possibly causing considerable damage. Furthermore, the planting of vetiver is an agronomical technique which does not involve any kind of pollution. Vetiver holds the soil, the fertilizers, etc. and does not allow the soil to be transported downstream, which would result in the impoverishment of the soil for farming purposes. The fact that the soil is not washed away and that the fertilizers are used in a more efficient way reduces the use and the waste of fertilizers and has important agricultural, ecological, economical and environmental repercussions.

A further indirect economic advantage of vetiver is the increase in the technical abilities of the farmer and the opening of new horizons and cultural awareness.

As was outlined in the paragraph "oil extraction" the vetiver farmer will be able to start small industrial and commercial enterprises which make the work more pleasant, especially for the young farmers. These activities should stimulate the young farmers so that they are less likely to give up farming. This stimulus, together with the use of vetiver in its formidable soil erosion control capacity, has important ecological, agronomical, social and political implications.

Direct advantages

The direct advantages from the increase in the farmer's income derive from:

the optimal use of the soil, and from soil conservation rather than erosion, which would otherwise lead to waste;

the increased fertility of the soil due to the reduced erosion of the fertile topsoil and consequent reduction in the use of fertilizers;

the opportunity to extract essential oil from the roots, with low-cost equipment, low energy consumption and easy maintenance;

the sale of handicraft products and manufactured handicraft articles based on the vetiver oil and plant;

the optimum use of time and workers on the farm: indeed, vetiver oil extraction and the production of handicraft products can be performed by female and elderly workers even during the periods of limited farming activity.

Planting vetiver is a simple, low-cost technique, lower in cost for example than other soil conservation techniques, artificial or natural. Vetiver oil (currently solely imported from third world countries) is sold at a price of approximately 110 ECU/kg in the European Union. The average return per hectare is estimated at least 40kg/ha with a surplus value of approximately 4,000 ECU/ha. To this amount the earnings that follow from the sale of handicraft. Finally, it must be considered that the large quantity of biomass produced per hectare (approximately 40 tons of dry matter per hectare) makes it possible to generate electricity on the farm itself.

All this makes vetiver a commodity with great potential and therefore of great interest for farmers in the European regions where a Mediterranean climate exists.

4 Chemical Constituents and Essential Oil Biogenesis in *Vetiveria Zizanioides*

Anand Akhila and Mumkum Rani

Central Institute of Medicinal and Aromatic Plants. Lucknow,
India 226 015

Introduction

Vetiveria zizanioides Linn. (Fam. Gramineae), named in Sanskrit as *reshira* or *sugandhmula* and in Hindi as *khus-khus* or *khas*, is native to India and the oil obtained from its roots has been known to Indians from the time of Vedas. During Mogul times, French traders introduced this plant to Bourbon Island in the Indian Ocean, and to the New World colonies of Louisiana and Haiti. Traditionally it was mainly produced in Java, the Reunion Islands and the Seychelles. But recently, a substantial portion of the total production has come from Haiti, Japan, Brazil and India. Two types of vetiver have been found in India – (i) flowering or seeding vetiver which grows wild in North India, and (ii) non-flowering or non-seeding vetiver which is cultivated in South India. The essential oils obtained from these two types of vetiver are different in their physico-chemical properties and chemical composition. The oil distilled from wild khus roots shows a characteristic high laevorotation while the oil from cultivated vetiver roots is dextrorotatory. Important centres of North Indian (wild) roots are Uttar Pradesh and Rajasthan. The cultivated variety is grown systematically in Kerala, Tamil Nadu, Karnataka and Andhra Pradesh.

Vetiver oil is one of the most complex mixtures of sesquiterpene alcohols and hydrocarbons, and also one of the most viscous oils with an extremely slow rate of volatility. It is used extensively for blending in cosmetics and in the soap industry and as a fixative in the perfumery industry prolonging the life of any composition to which it is added. In India it is used in herbal medicine as a carminative, stimulant and diaphoretic. The roots are used for the weaving of *tatis* (or *khuschiks*), screens, mats and fans. Several reviews of the chemistry of the essential oil and the utilization of vetiver have appeared in the past (Fuehrer, 1970, 1974; Akhila *et al.*, 1981; Bhatwadekar *et al.*, 1982).

Chemical Composition

The chemical examination of Indian vetiver oil has been the subject of investigation by many workers (Narain *et al.*, 1949; Zutsi *et al.*, 1956, 1957a, 1957b; Novotel'nova, 1958a, 1958b; Nigam *et al.*, 1959; Survey *et al.*, 1959; Nguyen-Trong-Ann, 1965; Manchanda *et al.*, 1968, 1970; Nanda *et al.*, 1970; Garnero, 1972; Masada and Sumido, 1979; Nair *et al.*, 1979) who have carried out fractional distillation of the oil and separated the constituents. So far over 75 sesquiterpenes have been isolated and some of these compounds have been synthesized chemically. Broadly, on the basis of structures, these compounds can be divided into following groups:-

A. Monocyclic sesquiterpenes
    (i)   Bisabolane (Table 4.1)
    (ii) Elemol (Table 4.1)
B. Bicyclic sesquiterpenes
    (i)   Eudesmane (Table 4.2)
    (ii) Nootkatane (Table 4.3)
    (iii) Spirane (Table 4.4)
    (iv) Cadinane (Table 4.5)
    (v) Cadinane norsesquiterpenes (Table 4.6)
C. Tricyclic sesquiterpenes
    (i)   Cedrane (Table 4.7)
    (ii) Zizaane and Prezizaane (Table 4.8)
D. Tetracyclic sesquiterpenes
    (i)   Cyclocopacamphane (Table 4.9)

Figure 4.1 shows all the basic skeletons found in vetiver oil and their possible biogenetic relationship. Biogenesis of vetiver sesquiterpenes has been discussed in several reviews (Nigam *et al.*, 1968; Andersen, 1970; Cordell, 1976) and attempts have also been made to synthesize these sesquiterpenes (Naegeli and Kaiser, 1972; Mehta *et al.*, 1973).

A. Monocyclic sesquiterpenes

(i) Bisabolane group

Isobisabolene (1), a sesquiterpene hydrocarbon was isolated from vetiver oil (Bharatpur variety) (Kalsi *et al.*, 1962) and the structure was assigned on the basis of spectroscopic and chemical studies. Later on, Vig *et al.* (1969) synthesized this compound by a Wittig reaction with methyltriphenyl phosphonium iodide on 4-(1-oxo-5-methyl-4-hexenyl)-cyclohexanone and later on (Vig *et al.*, 1971) by employing β-ketosulphoxides and their alkylated derivatives as key intermediates.

    β-Bisabolol (2) was isolated from the oil of *Vetiveria zizanioides* Stapf by Kaiser and Naegeli (1972). Mizrahi and Nigam (1969) reported the presence of dehydrocurcumene (4) in Reunion vetiver oil.

    Biosynthesis of the compounds of this group is initiated by the cyclisation of farnesyl pyrophosphate (FPP, Figure 4.2) in which an enzyme bond species (Enz) attacks C-7 followed by attack of $\Delta^{6,7}$ on C-1 facilitating the release of pyrophosphate. This ultimately generates the monocyclic bisabolane skeleton which by proton loss from different carbons, synthesizes compounds of this group *in vivo*.

(ii) Elemol group

Andersen (1970a) isolated elemol (3) from Reunion and Haiti vetiver oils while Homma *et al.* (1970) isolated it from Japanese vetiver oil.

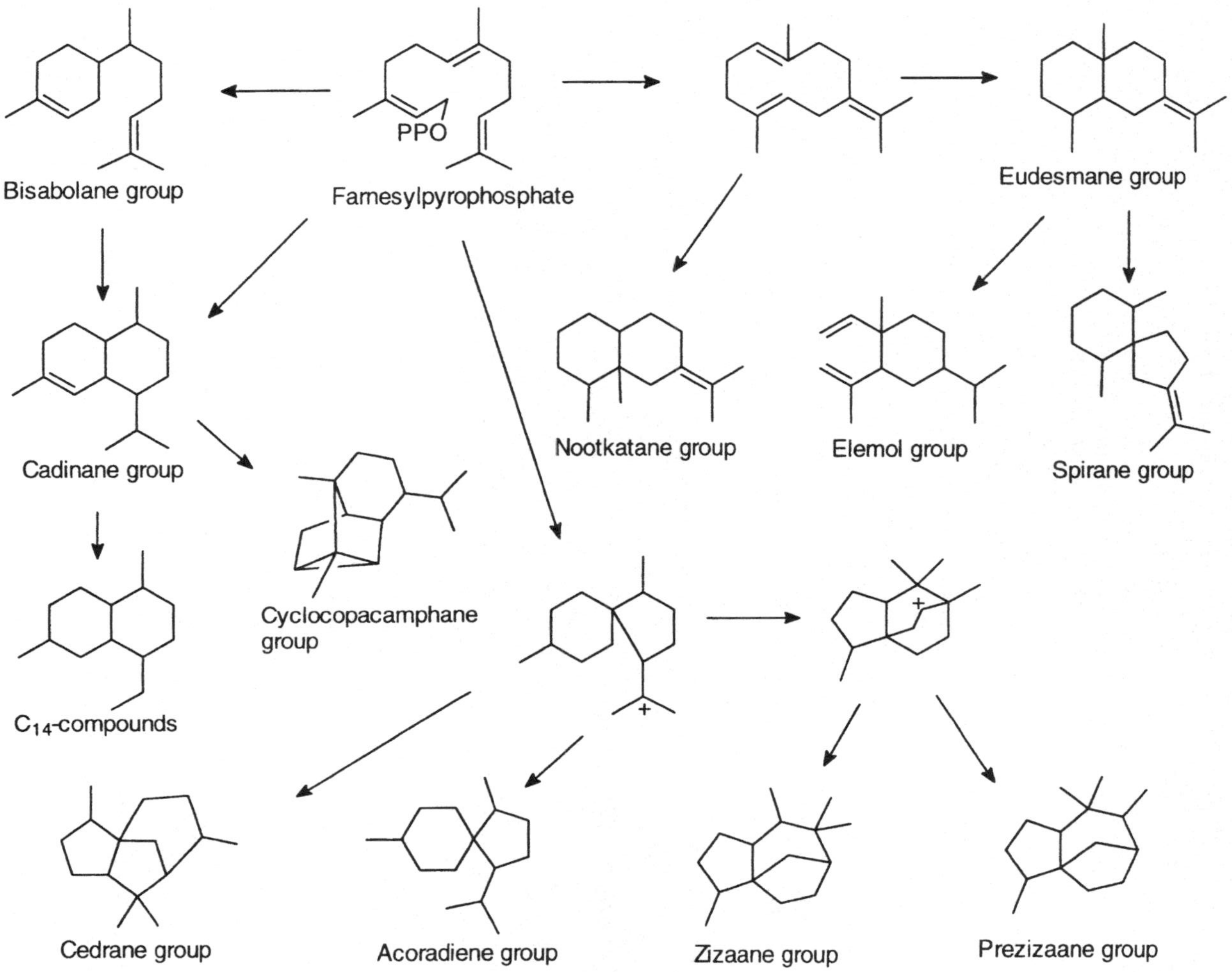

Figure 4.1 Possible biogenetic relationship amongst all the sesquiterpene skeletons present in the essential oil of *Vetiveria zizanioides*.

Table 4.1

Bisabolane and Elemol Group		Ref.

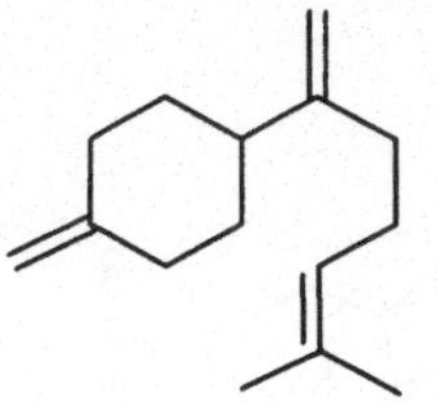

Isobisabolene (1)

$C_{15}H_{24}$; 204.35
b.p. 99–102°/8mm; $[\alpha]_D^{26}$ −47°
(c 4.6); n_D^{26} 1.4966; d_4^{26} 0.8859
Yield 3g/800g oil
Spectra: IR
Derivatives:
(i) Tetrahydroisobisabolene, b.p.
118°(bath)/8mm, $[\alpha]_D^{26}$ +28.8°
(c 4.05)); n_D^{26} 1.4780 (ii)
Hexahydrobisabolene, b.p. 92°
(bath)/0.3mm, $[\alpha]_D^{26}$ +8.9°
(c 5.2)); n_D^{26} 1.4573 (iii)
ozonolysis followed by treatment
with dimedone produced
formaldimethone, m.p. 183–186°
Synthesis

Kalsi *et al.*, 1962; Vig
et al., 1969, 1971

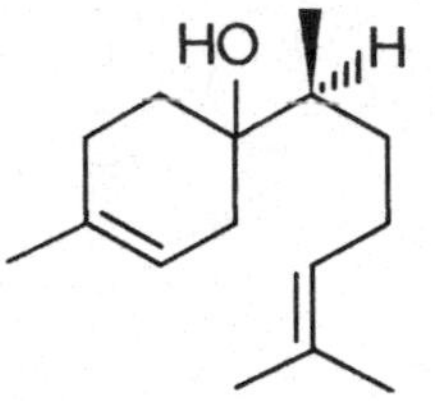

β-Bisabolol (2)

$C_{15}H_{26}O$; 222.37

Kaiser and Naegeli,
1972

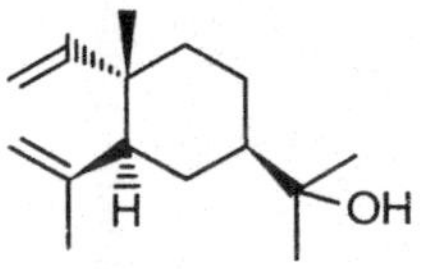

Elemol (3)

$C_{15}H_{26}O$; 222.37
m.p. 51–52°; $[\alpha]_D^{25}$ −4.5 (CHCl$_3$)
Spectra: IR, NMR
Derivative: *p*-Nitrobenzoate, m.p.
73–73°; $[\alpha]_D$ −7° (c 0.95,
CHCl$_3$); UV, ORD

Andersen, 1970a; Jones
and Sutherland, 1968

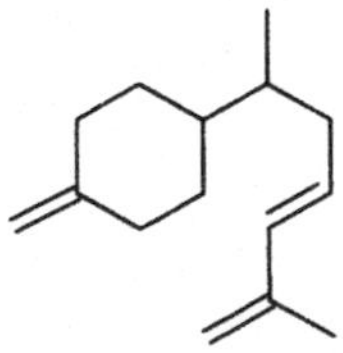

Dehydrocurcumene (4)

$C_{15}H_{24}$; 204.35
Spectra: UV, IR

Mizrahi and Nigam,
1969

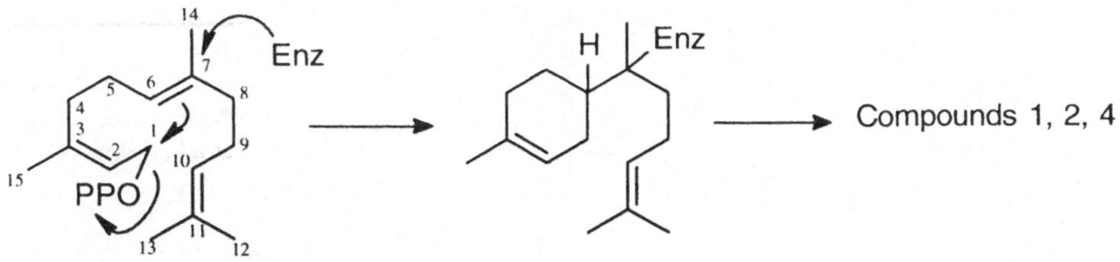

Figure 4.2 Cyclization of *cis*-farnesyl pyrophosphate (FPP) to monocyclic bisabolane compounds.

B. Bicyclic sesquiterpenes

(i) Eudesmane group

β-Eudesmol (5) co-occurs with elemol in Reunion and Haiti (Andersen, 1970a) and Japanese (Homma *et al.*, 1970) vetiver oils. 10-Epi-γ-eudesmol (6) had already been synthesized by Marshall and Pike in 1968 prior to its isolation from vetiver oil (Kaiser and Naegeli, 1972). It was assumed to be the precursor of the nootkatanes, α- and β-vetivane derivatives (Kaiser and Naegeli, 1972).

Laevojunenol (7), the first eudesmanic compound having the C(7) side chain α-oriented, was isolated from North Indian vetiver oil (Moosanagar variety) by Bhattacharyya *et al.* (1960). On selenium dehydrogenation it gave eudalene while on catalytic hydrogenation it gave the dihydro-alcohol and ozonolysis gave formaldehyde and a crystalline keto-alcohol, $C_{14}H_{24}O_2$, m.p. 43°, $[\alpha]_D$ −11.5°. The parent alcohol, its dihydro derivative and the C_{14}-keto alcohol were similar in all respects to junenol and its corresponding derivatives, except for the respective rotation values, which were equal in magnitude but opposite in sign. On the basis of these observations it was concluded to be the optical antipode of the dextrorotatory alcohol junenol which was isolated by Herout *et al.* (19??) from Juniper oil. It has also been reported to co-occur with zizaene (Andersen, 1970b) and its absolute configuration has been given by Shaligram *et al.* (1962).

Andersen *et al.* (1970d) reported vetiselinene (8) along with (-)-selina-4(14),7(11)-diene (9) and (-)-δ-selinene (10) in vetiver oil. Japanese vetiver oil (Homma *et al.*, 1970) was reported to contain vetiselinenol (11) while the investigation of Bharatpur variety of Indian vetiver oil (Karkhanis *et al.*, 1978) revealed the presence of vetiselinenol (11) and iso-vetiselinenol (12). The structure of vetiselinenol (11) was assigned on the basis of IR and ^{1}H NMR studies. Irradiation at δ 1.94 (=C−CH$_2$−) transformed the =CH− signal to a sharp singlet and the C=CH$_2$ signals became a pair of doublets (J = 1.6 Hz). A −CH(CH$_3$)CH$_2$OH grouping was suggested by the significant m/e 161 fragment in the mass spectrum. Furthermore, irradiation at δ 2.25 produced a simple AB pattern for the −CH$_2$OH grouping and also collapsed the CH$_3$ doublet at 1.01. The downfield position of this methine signal (H-11) suggested its allylic position. The formation of an antipode of 4,5α-H-eudesmane on hydrogenation of vetiselinene and the extreme upfield position of the angular methyl in vetiselinene confirmed a *trans* ring fusion and suggested 7(8)-position for the trisubstituted double bond.

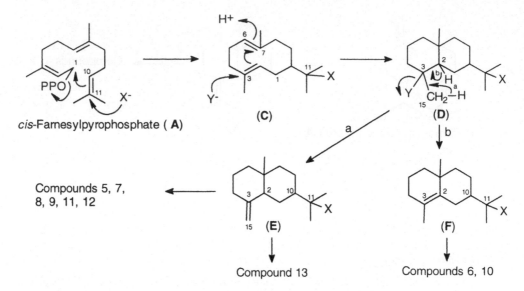

Figure 4.3 Biosynthesis of eudesmane group of compounds. X and Y denote enzymes (nucleophiles) or their biogenetic equivalents.

The ^{1}H NMR spectrum of isovetiselinenol (12) confirmed the position of hydroxyl group at C-2. It showed a widespread multiplet at 3.75. Of the various positions available for the attachment of the secondary −OH group in the vetiselinene skeleton, positions C-3, C-6 and C-9 being allylic, methine proton at one of these positions would appear at a lower field and the methine proton in none of these positions could appear as a multiplet. Similarly the methine proton at C-1 would appear as a triplet or at best as a quartet. Beside the observed signal at 3.75 there is a widespread multiplet, which showed the presence of more than two vicinal protons. This confirmed the position of an OH group at C-2.

A C$_{12}$-ketone, (+)-(6S, 10R)-6,10-dimethylbicyclo[4.4.0]dec-1-en-3-one (13) was isolated from Reunion vetiver oil. The structure was assigned using spectroscopic data and was confirmed by direct comparison with an authentic racemic sample. The absolute configuration was established by chemical correlation with (+)-α-eudesmol.

Biosynthesis of this group of compounds is initiated by already established biosynthetic mechanisms (Akhila *et al.*, 1989) of cyclization of FPP (Figure 4.3). An enzyme or its biogenetic equivalent (X⁻) attacks at C-11 and $\Delta^{10,11}$ opens to form a bond with C-1 to give rise to a biogenetically equivalent intermediate (C). Another enzyme Y⁻ further initiates cyclization by attack on C-3 of C and future losses of protons from C-15 and C-2 produce intermediates **E** and **F** respectively which are the likely penultimate precursors of all the compounds of this group.

(ii) Nootkatane group

Andersen *et al.* (1970d) reported valencene (14), nootkatene (15) and β- & γ-vetivenenes (16, 17). The structures were assigned by ^{1}H NMR. In the case of valencene (14) irradiation at δ 1.71 transformed the signal at δ 4.67 to a sharp singlet and *vice-versa*.

Table 4.2

Eudesmane Group		Ref.
 β-Eudesmol (**5**)	$C_{15}H_{26}O$; 222.37 Waxy solid; $[\alpha]_D$ +61° **Spectra:** IR, NMR **Derivative:** Formic acid treatment afforded (+)-δ-selinene	Andersen, 1970a
 10-Epi-γ-eudesmol (**6**)	$C_{15}H_{26}O$; 222.37 b.p. 105–110° (bath)/0.1 mm; n^{20}_D 1.5110 **NMR:** $\delta_{TMS}^{CCl_4}$ 2.25 (OH), 1.64 (vinyl CH_3), 1.19, 1.13 and 1.07 (three CH_3)	Kaiser and Naegeli, 1972; Maurer *et al.*, 1972
 Laevojunenol (**7**)	$C_{15}H_{26}O$; 222.37 m.p. 65° (crystallized from petroleum ether) $[\alpha]_D$ −57° (c 1.18) **Spectra:** IR, NMR **Derivatives:** (i) Eudalene on Se dehydrogenation (ii) Dihydro-laevojunenol, m.p. 115°, $[\alpha]_D$ ±0° (c 2.6) (iii) Ozonolysis: ketoalcohol, m.p. 43°, $[\alpha]_D$ −11.5° (c 2.00)	Andersen, 1970b; Bhattacharyya *et al.*, 1960; Shaligram *et al.*, 1962
 Vetiselinene (**8**)	$C_{15}H_{24}$; 204.35 $[\alpha]_D^{27}$ −20° (c 2.3) **Spectra:** IR, ^{1}H NMR ^{1}**H NMR:** δ (CHCl₃) 0.64 (s, 3H), 1.14 (d, J = 6.5 Hz, 6H), 4.52, 4.70 (br. S, 1H each, vinyl), 5.24 (br. S, 1H, >C=C$\underline{H}$−).	Andersen *et al.*, 1970d; Andersen, 1970a; Homma *et al.*, 1970; Karkhanis *et al.*, 1978
 (-)-Selina-4(14), 7(11)-diene (**9**)	$C_{15}H_{24}$; 204.35	Andersen *et al.*, 1970d

Table 4.2 (cont'd)

Eudesmane Group		Ref.

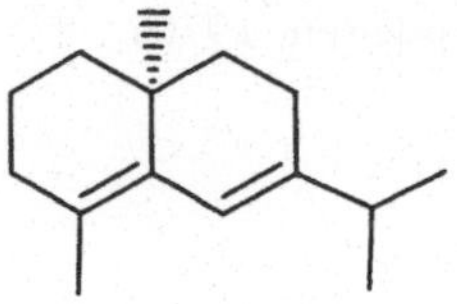

(-)-δ-Selinene (10)

$C_{15}H_{24}$; 204.35

Andersen *et al.*, 1970d

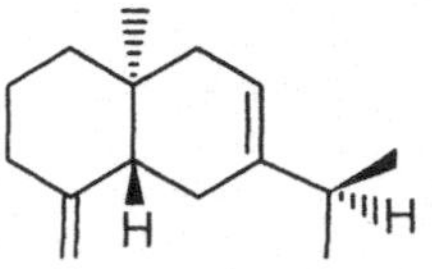

Vetiselinenol (11)

$C_{15}H_{24}O$; 220.35
Colourless oil; $[\alpha]_D^{18}$ −18.1°
(CHCl$_3$), $[\alpha]_D^{27}$ −30° (c 3.1)
Spectra: UV, IR, ^{1}H NMR
^{1}**H NMR:** δ (TMS, CDCl$_3$) 0.69
(3H, s, CH$_3$), 1.01 (3H, d, J =
6.8 Hz, CH$_3$), 1.94 (6H, W$_h$
~2.5 Hz), 2.0–2.5 (2H, m,
allylic protons), 3.52 (apparent
d, 6.2 Hz, CH$_2$OH), 4.545 and
4.78 (C=CH$_2$), and 5.43
(W$_h$ ~8 Hz, =CH−).
Derivative: (i) Ester with
CH$_2$N$_2$, IR, ^{1}H NMR (ii)
Vetiselinene (iii) (+)selinane, IR,
MS, ^{1}H NMR (iv) monoepoxide,
m.p. 144–145°, IR, NMR

Andersen, 1970a;
Homma *et al.*, 1970;
Karkhanis *et al.*, 1978

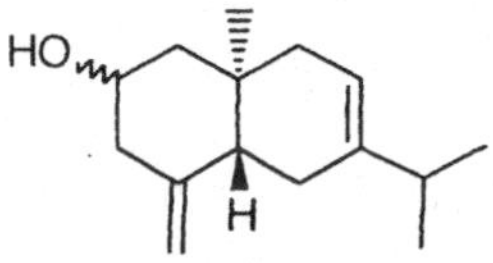

Iso-Vetiselinenol (12)

$C_{15}H_{24}O$; 220.35
b.p. 105–107°(bath)/0.2 mm;
$[\alpha]_D^{25}$ −11.2° (c 3.3)
Spectra: IR, MS, ^{1}H NMR
^{1}**H NMR:** δ (TMS, CDCl$_3$) 0.66
(3H, s, CH$_3$), 1.01 (6H, d, J =
7 Hz, gem methyls), 1.61 (1H,
disappears after D$_2$O exchange,
−OH), 1.91 (6H, s, W$_{1/2}$ = 3 Hz,
allylic protons), 2.1–2.66 (2H,
m, allylic protons), 3.75 (1H, m,
>CHOH), 4.65 and 4.85 (1H
each, >C=CH$_2$) and 5.25 (1H,
br s, W$_{1/2}$ = 8 Hz, >C=CH−)

Karkhanis *et al.*, 1978

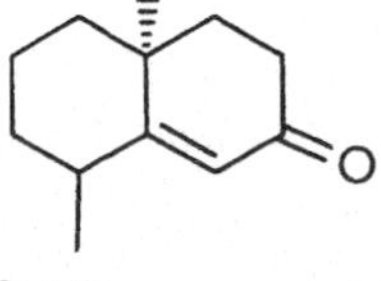

C-12 compound (13)

$C_{12}H_{18}O$; 178.27
$[\alpha]_D^{20}$ +201.7° (c 2.0)
Spectra: UV, IR, MS, ^{1}H NMR
^{1}**H NMR:** δ (TMS, CCl$_4$) 1.06
(3H, d, J = 6 Hz), 1.25 (3H, s),
5.61 (1H, d, J = 2Hz)
Derivative: Dihydro-compound,
$[\alpha]_D^{20}$ +39° (c 1.0), IR, Mass, ^{1}H
NMR

Maurer *et al.*, 1972

The signal at 5.34 was observed as a sharp singlet by irradiation at δ 1.96 and the fine structure of the allylic proton region (1.75–2.40) became somewhat clearer by irradiation at 5.34. All these observations support the given structure of valencene.

The NMR spectrum of β-vetivenene (16) supported an unambiguous assignment of the position of the double bond in the nootkatane skeleton (established by hydrogenation). β-Vetivenene showed the negative trending RD curve analogous to that of nootkatene and also showed a nearly identical vinyl-H region of the NMR.

^{1}H NMR spectra of γ-vetivenene (17) showed signals at δ 0.92, 1.04, 1.915, 2.32, 4.91, 5.03, 5.38 and 5.95 ppm. The downfield position of the resonances for the isopropenyl group and the singlet (W_h ~3 Hz) vinyl-H at δ 5.95 indicate that these two groups (i.e. isopropenyl and vinyl) are in conjugation. The position of remaining double bond was fixed at C-9(10) on the basis of ORD and hydrogenation results.

Isovalencenol (18) was isolated by Karkhanis *et al.* (1978) from North Indian vetiver oil as shining white needles. The IR and ^{1}H NMR studies showed it to possess a nootkatane skeleton. An isomer of this compound, bicyclovetivenol, (19) had been isolated from Japanese vetiver oil (Takahashi, 1968), and was a liquid, b.p. 150°/0.02 mm, $[\alpha]_D$ +122.3°. Other properties were the same.

Isovalencenic acid (20), a minor acidic constituent was isolated from vetiver oil of Japanese origin. The geometrical disposition of carboxylic acid has been confirmed by its conversion to a γ-lactone. The methyl ester of isovalencenic acid on reduction with LiAlH$_4$ gave a solid primary alcohol, identical in all respect with the crystalline isovalencenol (18).

α-Vetivone (21), also known as isonootkatone, was obtained from Haiti oil of Vetiver via separation of the ketonic components with Girard's T reagent and its structure has been established (Marshall and Andersen, 1967). α- and β-vetivones were also characterized from the Reunion oil by GLC (Nigam and Levi, 1962). Total synthesis of 21 was achieved from 2-isopropylidene-1, 3-propanediol in about 14 steps (Marshall and Warne Jr., 1971). The key step involved annelation of 4-isopropylidene-2-carbomethoxycyclohexanone with *trans*-3-penten-2-one to give the bicyclic enone with cis-related CH$_3$ and COOCH$_3$ substituents. 21 has also been reported from North Indian Vetiver oil (Kirtany and Paknikar, 1971). Despite an extraordinary resemblance to β-vetivone (26) in its chemical reactions and in most of its physical properties, the significant facts upon which the structure of 21 was based were (a) the presence of an isopropylidene group, (b) the similarity of the UV spectrum, (c) a parallel behaviour in hydrogenation and (d) the formation of vetivazulene on dehydrogenation (Endo and Mayo, 1967). A method for the stereoselective construction of 21 and nootkatone (22) has been demonstrated from bicyclo[2.2.2]octyl esters and 1-methoxy-3,4,5-trimethyl- cyclohexa-1,4-diene (Dastur, 1974). A C$_{12}$-ketone, (+)-(1S,10R)-1,10-dimethylbicyclo[4.4.0]dec-6-en-3-one (23) was isolated from Reunion vetiver oil. The structure and absolute configuration of 23 were established by a four step synthesis from 21 (Maurer *et al.*, 1972).

Biogenetic rearrangement of the eudesmane skeleton to the nootkatane skeleton has been demonstrated in Figure 4.4 and Akhila *et al.* (1989) have conducted several radio-labelled experiments on the compounds of this group to provide unequivocal evidence of a 1,2-methyl shift and other proton losses. During this rearrangement, C-14 methyl shifts to C-2 of the eudesmane skeleton (**D**) to form **G** which is the precursor of all the nootkatane group compounds.

Table 4.3

Nootkatane Group		Ref.

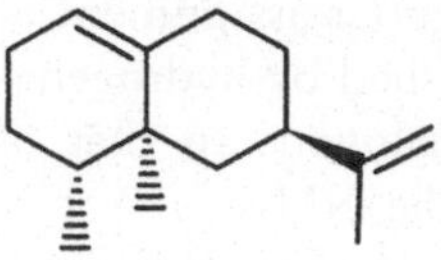

Valencene (14)

$C_{15}H_{24}$; 204.35
$[\alpha]_D$ +190°; d_4^{21} 0.9199; n_D^{20} 1.5041
Spectra: ^{1}H NMR
^{1}**H NMR:** δ 0.86 (3H, d, J = 6 Hz, CH_3), 0.95 (3H, s, CH_3), 1.71 (3H, t, J = 1.0 Hz, allylic methyl), 4.67 (2H, t, J = 1.0 Hz, vinyl protons), 5.34 (1H, br m, >C=C$\underline{H}$–),

Andersen *et al.*, 1970d; Ishida *et al.*, 1970

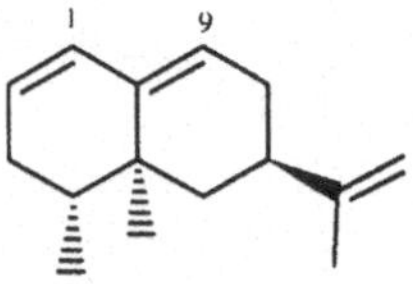

Nootkatene (15)

$C_{15}H_{22}$; 202.33
$[\alpha]_D$ −160°
Spectra: UV, IR, RD, ^{1}H NMR
^{1}**H NMR:** δ (TMS, $CDCl_3$) 0.89 (3H, d, CH_3), 0.92 (3H, s, CH_3), 1.75 (3H, s, viny-CH_3), 4.73 (2H, vinyl protons), 5.4 (1H, m, H-9), 5.56 (1H, m, H-2) and 5.97 (1H, d, J = 10 Hz, H-1).

Andersen *et al.*, 1970d

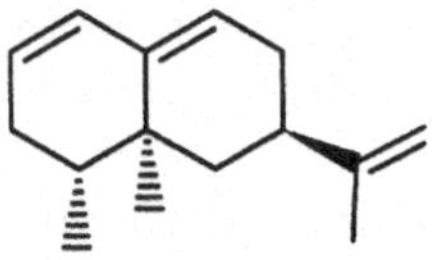

β-Vetivenene (16)

$C_{15}H_{22}$; 202.33
$[\alpha]_D$ −160°; n_D^{20} 1.5361
Spectra: UV, ^{1}H NMR
^{1}**H NMR:** δ 2.87 (2H, H-8), 5.4 (1H, H-9), 5.6 (1H, H-2), 5.94 (1H, H-1)

Andersen *et al.*, 1970d

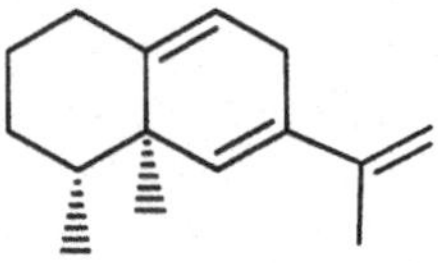

γ-Vetivenene (17)

$C_{15}H_{22}$; 202.33
Spectra: UV, IR, ^{1}H NMR
^{1}**H NMR:** δ (TMS, C_6D_6) 0.92 (3H, d, J = 6.8 Hz, CH_3), 1.04 (3H, s, CH_3), 1.915 (3H, vinyl-CH_3), 2.32 (2H, =C–C$\underline{H}_2$), 4.91 and 5.03 (1H each, >C=C$\underline{H}_2$), 5.38 (1H, ~triplet, J ≅ 3.5 Hz, =CH–) and 5.95 (1H, s, =CH–).

Andersen *et al.*, 1970d

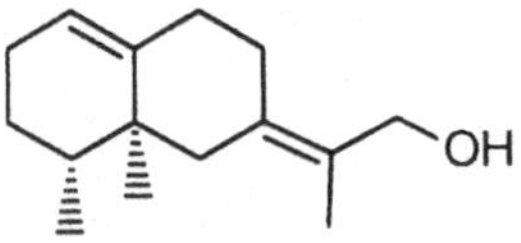

Isovalencenol (18)

$C_{15}H_{24}O$; 220.35
m.p. 89–90° (Shining white needles); $[\alpha]_D$ +162.8°
Spectra: IR, ^{1}H NMR
^{1}**H NMR:** δ 0.82 (3H, s, CH_3), 0.93 (3H, d, J = 5.5, >CH–C$\underline{H}_3$), 1.33 (1H, disappears on D_2O exchange, –OH), 1.75 (3H, s, =C–CH_3), 4.12 (2H, s, >C=C–CH_2OH) and 5.3 (1H, br s, >C=CH–).
Derivatives: (i) Tetrahydrovalencenol, IR, ^{1}H NMR, (ii) Ozonolysis : Hydroxyacetone,

Karkhanis *et al.*, 1978

Table 4.3 (cont'd)

Nootkatane Group		Ref.

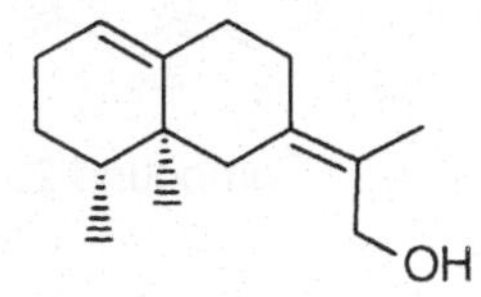

Bicyclovetivenol (**19**)

$C_{15}H_{24}O$; 220.35
b.p. 150°/0.02 mm, $[\alpha]_D$ +122.3

Karkhanis *et al.*, 1978

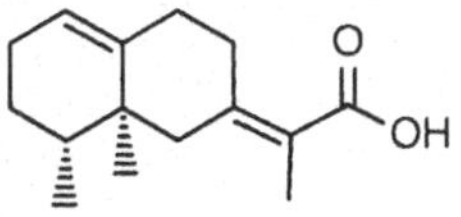

Isovalencenic
acid (**20**)

$C_{15}H_{22}O_2$; 234.33
m.p. 135–137°
Spectra: UV, IR, ^{1}H NMR
^{1}**H NMR**: δ TMS, CCl$_4$) 0.86 (3H,
s, CH$_3$), 0.93 (3H, d, J = 6, CH$_3$),
1.93 (3H, s, =C–CH$_3$), 5.32 (1H,
m, H-1).
Derivatives: (i) Methyl ester, $[\alpha]_D^{22}$
+136.6°, UV, IR, ^{1}H NMR, (ii)
Di- and tetrahydro-derivative of
methyl ester, UV, IR, ^{1}H NMR,

Hanayama *et al.*, 1968;
Karkhanis *et al.*, 1978

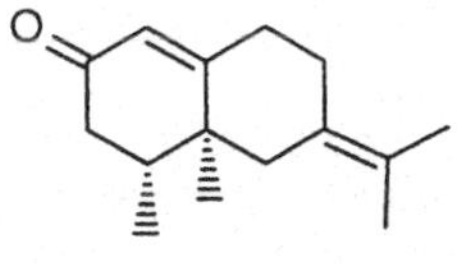

α-Vetivone (**21**)

$C_{15}H_{22}O$; 218.33
b.p. 110°/0.015 mm

Dastur, 1974; Endo
and Mayo, 1967;
Kirtany *et al.*, 1971;
Marshall *et al.*, 1967;
Marshall *et al.*, 1971;
Nigam and Levi, 1962;
Pfau *et al.*, 1939

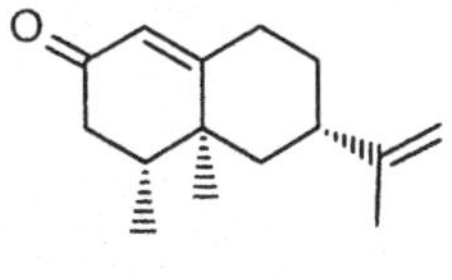

Nootkatone (**22**)

$C_{15}H_{22}O$; 218.33
^{1}**H NMR**: δ (CCl$_4$), 5.62 (1H, s,
olefinic proton), 4.68 (2H, s, vinyl
protons), 1.72 3H, s, olefinic
methyl), 1.11 (3H, s, CH$_3$), 0.95
(3H, d, J = 6 Hz, CH$_3$).
Synthesis

Dastur, 1974; Marshall
and Ruden, 1971

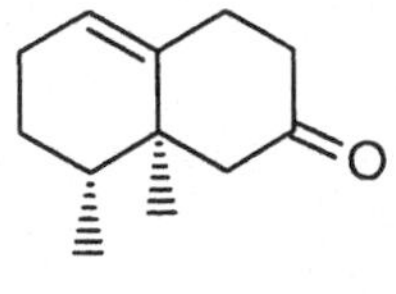

C-12 Compound (**23**)

$C_{12}H_{18}O$; 178.27
$[\alpha]_D^{20}$ +101.2°
Spectra: UV, IR, ^{1}H NMR, MS
^{1}**H NMR**: 0.90 (3H, s, CH$_3$), 0.91
(3H, d, J = 6 Hz, CH$_3$), 5.45 (1H,
m, olefinic proton)

Maurer *et al.*, 1972

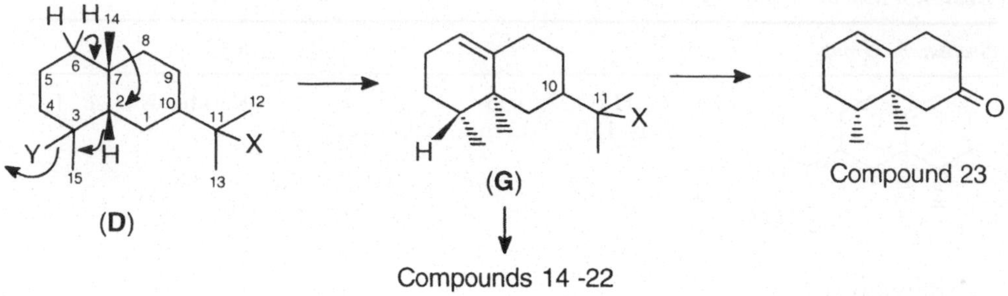

Figure 4.4 Biogenetic rearrangement of eudesmane skeleton to nootkatane group of compounds.

(iii) Spirane group

The NMR spectrum (Andersen *et al.*, 1970d) of β-vetispirene (24) showed a conjugated butadiene system. The downfield signal at δ 5.92 confirmed an internal vinyl-hydrogen on the conjugated system (H-2). The 10.0 Hz coupling showed a *cis* double bond while the smaller coupling (2.6 Hz) was due to an axial allylic hydrogen at δ 2.4. Irradiation at δ 2.4 produced a doublet at δ 5.92 (J = 10 Hz) and reduced the complex multiplet at δ 5.50 (due of H-1) to a doublet of doublets (J = 10, 5 Hz). Irradiation at δ 2.0 left H-2 unchanged but reduced the H-1 resonance to a doublet of doublets (J = 10, ~2 Hz). In α-vetispirene (25), the isopropenyl group (CH$_3$ 1.91, C=CH$_2$ 4.85, downfield from the usual position) and the singlet vinyl-H (δ 5.46) was assigned to be in the conjugation. The absence of significant coupling showed the absence of any allylic hydrogens at the H-terminus of the double bond. β-Vetivone (26) was isolated from Haiti and Reunion varieties of vetiver (Andersen, 1970a) and its total synthesis has also been reported (Dauben and Hart, 1975; Deighton *et al.*, 1975). A confirmation of the revised structure of 26 was secured upon completion of total synthesis of its racemic form by a stereoselective pathway (Marshall and Johnson, 1970b). A C$_{12}$ compound has been reported (personal communication). Synthesis of hinesol (28) has been achieved by various workers (Marshall and Brady, 1969; Yamada *et al.*, 1973; Buddhsukh and Magnus, 1975; Dauben and Hart, 1975). The isolation and structure determination of acoradiene 29 and 30 has been achieved along with several minor and biogenetically interesting sesquiterpenes of vetiver oil (Paknikar *et al.*, 1975).

The eudesmane skeleton **D** appears to be the intermediate in the biosynthetic sequence of spirane compounds shown in Figure 4.5. Though systematic studies using radio-active precursors have not so far been reported it can be safely suggested that the bond between C-7 and C-8 rearranges to C-2 in skeleton **D** and intermediate (**H**) is formed and then converted to this group of compounds.

(iv) Cadinane group

The isolation of khusinol (31) from the laevorotatory vetiver oil (*V. zizanioides* Linn.) from Bharatpur and Biswan area of North India was first reported in 1963 by Rao *et al.* (1963) who assigned the adjacent structure to this sesquiterpene alcohol on the

Table 4.4

Spirane Group		Ref.

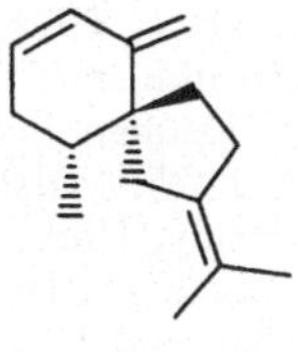

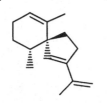

β-Vetispirene or
β-isovetivenene (**24**)

$C_{15}H_{22}$; 202.33
$[\alpha]_D$ −90°
Spectra: UV, IR, ^{1}H NMR
^{1}**H NMR:** δ (TMS, CCl_4) 0.83
(3H, d, J = 6.6 Hz, CH_3), 1.62 &
1.67 (3H each, C=C−Me_2), 4.69 &
4.77 (1H each, C=CH_2), 5.50 (1H,
complex multiplet, vinyl-H), 5.92
(1H, dd, J = 10.0 & 2.6 Hz, vinyl-H).

Andersen *et al.*, 1970d

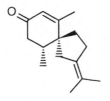

α-Vetispirene (**25**)

$C_{15}H_{22}$; 202.33
$[\alpha]_D$ +220°; $[\alpha]_{200}$ +2300°
Spectra: UV, IR, ^{1}H NMR
^{1}**H NMR:** δ (TMS, CCl_4) 0.867
(3H, d, J = 5.6 Hz, CH_3), 1.544
(3H, ~d, vinyl-CH_3), 1.91 (3H, s,
vinyl-CH_3), ~2.54 (a 10 line
multiplet due to two hydrogens),
4.85 (2H, apparent quartet, S = 0.5
Hz, C=CH_2), 5.33 (1H, tq, J_q = 1.4,
J_t = 4 Hz, vinyl-H), 5.46 (1H, s,
W_h = 3.2, vinyl-H).

Andersen *et al.*, 1970d

β-Vetivone (**26**)

$C_{15}H_{22}O$; 218.33
m.p. 46–48°; $[\alpha]_D^{26}$ −23.6°
($CHCl_3$, c 1.0)
Spectra: IR, UV, ^{1}H NMR
^{1}**H NMR:** δ (TMS, CCl_4) 0.99 (3H,
d, J = 6 Hz, CH_3), 1.44 (6H, gem-
methyls), 1.90 (2H, CH_2CO), 1.93
(3H, C=C−CH_3), 5.60 (1H, $W_{1/2}$
= 3.3 Hz, =CH−CO).
Derivative: 2,4-
dinitrophenylhydrazone, m.p.
182–184°
Synthesis

Private communication

Andersen, 1970a;
Buchi *et al.*, 1976;
Dauben and Hart,
1975; Deighton *et al.*,
1975; Kirtany *et al.*,
1971; Marshall *et al.*,
1969, 1970b; Nigam
and Levi, 1962; Pfau
et al., 1939; Pfau,
1940

C-12 Compound (**27**)

Table 4.4 (cont'd)

Spirane Group		Ref.
Hinesol (**28**)	C$_{15}$H$_{26}$O; 222.37 [α]$_D^{25}$ −40° **Synthesis**	Buchi *et al.*, 1976; Buddhsukh *et al.*, 1975; Dauben and Hart, 1975; Marshall and Brady, 1969; Yamada *et al.*, 1973; Marshall and Brady, 1970
Acoradiene (**29**)	C$_{15}$H$_{24}$; 204.35 [α]$_D$ −3.8° (CHCl$_3$, c 0.522) **Spectra:** IR, ^{1}H NMR, MS ^{1}**H NMR:** δ (TMS, CDCl$_3$) 0.80 (3H, d, J = 6.5 Hz, CH$_3$), 0.91 (3H, d, J = 6.5 Hz, CH$_3$), 1.50–1.70 (6H, m, vinyl methyls), 5.30 (1H, m, vinyl-H), 5.40 (1H, m, vinyl-H).	Kaiser and Naegeli, 1972; Paknikar *et al.*, 1975
Acoradiene (**30**)	C$_{15}$H$_{24}$; 204.35 [α]$_D$ +34.4° (CHCl$_3$, c 0.244) **Spectra:** IR, ^{1}H NMR, MS ^{1}**H NMR:** δ (TMS, CDCl$_3$) 0.87 (3H, d, J = 6.5 Hz, CH$_3$), 1.01 (6H, d, J = 6.5 Hz, 2 x CH$_3$), 1.62 (3H, m, CH$_3$), 5.25–5.50 (2H, m, vinyl protons).	Kaiser and Naegeli, 1972; Paknikar *et al.*, 1975

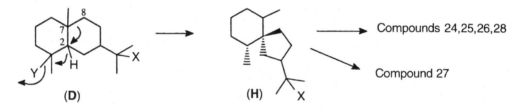

Figure 4.5 1, 2-Shifts involved in biogenesis of spirane group of compounds from eudesmane compounds

basis of chemical evidence and, apparently, without the benefit of NMR spectroscopy. Later on, a revised structure (**31**) was proposed for this compound on the basis of further chemical evidence and high resolution NMR studies (Tirodkar *et al.*, 1969).

The NMR signals in epikhusinol (**32**) are strikingly similar to those of khusinol (Kalsi *et al.*, 1972), the only significant difference being in the position and nature of the C<u>H</u>OH proton. In khusinol this appears at 4.06 as a broad multiplet whereas in epikhusinol this appears at a lower field (4.28) as a narrow multiplet. This clearly shows the axial orientation of the hydroxyl group in epikhusinol whereas here is an equatorial configuration in khusinol (31).

Khusol (33), a crystalline primary sesquiterpene alcohol belonging to the antipodal group of cadenenic alcohols was isolated (Kalsi et al., 1963) from North Indian vetiver oil. The structure and absolute configuration was assigned on the basis of chemical degradation and IR and ^{1}H NMR spectroscopy. γ-Cadinene (34) (Trivedi et al., 1971) and γ_2-Cadinene (35) (Kartha et al., 1963) were isolated from North Indian vetiver oil. The structure and absolute configuration of 35 has been established on the basis of formation of (+)-cadinene dihydrochloride on treatment of 35 with hydrogen chloride. Of the nine possible cadinene isomers β-, γ-, γ_1-, ϵ and δ-isomers are capable of forming crystalline dihydrochloride derivatives. A total synthesis of ($\pm$)-γ_2-cadinene has been reported from keto-enol ether which was earlier prepared in 8 steps (Kelly and Eber, 1970).

Investigation of the alcoholic fraction of North Indian vetiver oil led to the isolation of a crystalline epoxy alcohol named khusinol oxide (36) because of its structural relation with khusinol (Seshadri et al., 1967). The structure was deduced by ^{1}H NMR spectroscopy. It showed signals at 9.32, 9.19, 9.13 and 9.02 τ (6H), due to an isopropyl group; a signal at 8.35 τ (3H) was due to a methyl group on a double bond; a doublet centred at 7.39 τ (J = 4 c/s, 2H) showed the presence of the protons on the epoxide ring; a broad triplet at 6.38 τ (1H) was due to the proton on the carbon atom bearing the secondary OH group; a signal at 5.65 τ (1H) could be attributed to the OH proton, which disappeared on D_2O exchange; and a broad signal at 4.55 τ (1H) was attributed to the proton on the trisubstituted olefinic linkage. The β-orientation of the epoxy group was assigned from the β-orientation of the OH group in khusinodiol (39) which was obtained by reduction of 36 with LAH (Seshadri et al., 1967).

(+)-Amorphene or zizanene (37) has been reported to co-occur with laevojunenol (7) (Andersen, 1970b) in North Indian variety. α-Calacorene (38) was found in Reunion vetiver oil along with two other aromatic sesquiterpene hydrocarbons, one of which was dehydrocurcumene (4) whereas the other was a new cadalene-type hydrocarbon (Mizrahi and Nigam, 1969).

Further processing of the alcoholic fraction of the North Indian vetiver oil freed from the major sesquiterpenes led to the isolation of khusinodiol (39), (+)-α-cadinol (40) and cadina-4α,10β-diol (41) (Kalsi et al., 1979a). In these compounds the β-orientation of the OH group at C-10 seems to have a biogenetic relationship with khusinol oxide (36) in which the epoxy group is also β-oriented.

Vetidiol (42) has been isolated (Kalsi and Talwar, 1987) from the benzene fraction of North Indian vetiver oil. The ^{1}H NMR data, coupled with the formation of cadalene on dehydrogenation, showed that one hydroxyl group in 42 is axial while the other is equatorial located at C-5. The allylic nature and the location of the β-axial OH group at C-8 was proved by the MnO_2 oxidation of 42 to the hydroxy ketone in which the exocyclic double bond is in conjugation with the keto group [(^{1}H NMR, δ 5.8 and 6.2, 1H each br s) and the presence of a broadened multiplet at 4.15 (1H, $W_{1/2}$ 20 Hz, $-C\underline{H}OH$)].

Attempts to isolate veticadinol (43) from Congo vetiver oil were unsuccessful (Chiurdoglu and Delesemme, 1961). However, dehydration of 43 afforded veticadinene which was isolated from the alcoholic fraction of the oil. The structure of veticadinene was proved on the basis of degradation products. Accordingly, the structure of 43 was established. The position of the methylene group in 43 was also established

Table 4.5

Cadinane Group		Ref.
 Khusinol (**31**)	$C_{15}H_{24}O$; 220.35 m.p. 87°, 115.5–116.5°; $[\alpha]_D^{25}$ 174.4° (c 5.03) **Spectra:** IR, ^1H NMR ^1H **NMR:** δ 1.77 (3H, s, vinylic CH$_3$), 4.15 (1H, m, $W_{1/2}$ = 20 Hz, C<u>H</u>OH), 4.50 & 4.62 (1H each, s, =CH$_2$), 5.57 (1H, s, vinylic proton). **Derivatives:** (i) Acetate, b.p. 150° (bath)/0.5 mm; $[\alpha]_D^{28}$ −168.4° (c 6.855); n_D^{28} 1.4999, IR (ii) formaldimethane, m.p. 189° (iii) Ketone, b.p. 155–160° (bath)/2.5 mm, $[\alpha]_D^{25}$ +49.86° (c 5.234); n_D^{27} 1.5315, UV, IR (iv) Dihydro-khusinol, m.p. 111°, $[\alpha]_D^{25}$ 12.15°, IR (v) Tetrahydrokhusinol, m.p. 93–94°, $[\alpha]_D^{24}$ 20.84°, IR (vi) (-)-γ-Cadinene, b.p. 130–133° (bath)/4.3 mm, $[\alpha]_D^{24}$ −153° (c 2.476), n_D^{23} 1.5083, d_D^{24} 0.9189, IR (vii) Methyl ether, $[\alpha]_D^{28}$ −133° (c 3.4), IR, ^1H NMR	Agarwal *et al.*, 1985; Kelly and Eber, 1972; Rao *et al.*, 1963; Tirodkar *et al.*, 1969; Kohli, 1976b
 Epikhusinol (**32**)	$C_{15}H_{24}O$; 220.35 b.p. 150° (bath)/1.5 mm; $[\alpha]_D^{25}$ −20° **Spectra:** UV, IR, ^1H NMR ^1H **NMR:** δ 0.75 & 0.90 (3H each, d, = 7 Hz, gem-methyls), 1.7 (3H, s, vinyl-methyl), 1.8 (1H, exchangeable with D$_2$O, OH), 4.3 (1H, m, C<u>H</u>OH), 4.8 (2H, >C=CH$_2$), 5.6 (1H, m, =CH). **Derivatives:** (i) Tetrahydroepikhusinol, (ii) Epikhusinol acetate, IR **Synthesis**	Kalsi *et al.*, 1972; Kohli and Badaisha, 1985; Kohli, 1977
 Khusol (**33**)	$C_{15}H_{24}O$; 220.35 m.p. 101–102°; $[\alpha]_D^{25}$ −137° (c 2.9) **Spectra:** IR, ^1H NMR **Derivatives:** (i) Acetate, b.p. 155° (bath)/0.7 mm, $[\alpha]_D^{24}$ −115° (c 2.0), n_D^{27} 1.5060, (ii) Dihydrokhusol, b.p. 155° (bath)/1.5 mm, $[\alpha]_D^{27}$ +60° (c 1.7), n_D^{27} 1.5080 (iii) Tetrahydrokhusol, b.p. 140° (bath)/0.5 mm, $[\alpha]_D^{27}$ +34° (c 2.0), n_D^{28} 1.4951 (iv) Aldehyde, b.p. 135° (bath)/1.5 mm, $[\alpha]_D^{27}$ −165° (c 2.5), n_D^{27} 1.5172.	Kalsi *et al.*, 1963; Kohli, 1976a

Table 4.5 (cont'd)

Cadinane Group		Ref.

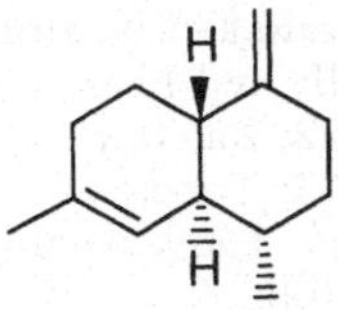

(-)-γ-Cadinene (**34**)

$C_{15}H_{24}$; 204.35
b.p. 100° (bath)/0.5 mm; $[\alpha]_D^{26}$
−145° (c 3.8); n_D^{31} 1.5060; d_4^{24}
0.9182
Spectra: IR
Derivative: Dihydro(-)-γ-cadinene,
b.p. 90°(bath)/0.8 mm; $[\alpha]_D^{27}$
+70° (c 0.9); n_D^{27} 1.4900

Tomita, 1971

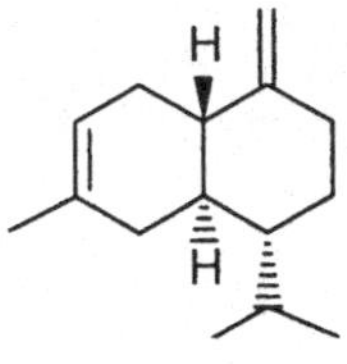

(-)-γ₂-Cadinene (**35**)

$C_{15}H_{24}$; 204.35
b.p. 115°/3 mm; $[\alpha]_D$ −40°; n_D^{28}
1.5051; d_4^{30} 0.9168
Spectra: IR
Derivatives: (i) Tetrahydro-γ₂-
cadinene, b.p. 133° (bath)/9 mm;
d_4^{30} 0.8789; n_D^{28} 1.4800 (ii)(+)-
Cadinene dihydrochloride, m.p.
117°, $[\alpha]_D$ +37.5° (iii) Dimedone
derivative, m.p. 188°
Synthesis

Kartha *et al.*, 1963;
Kelly and Eber,
1970

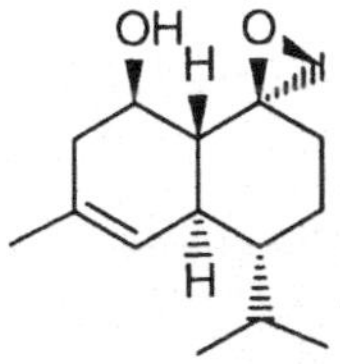

Khusinol oxide (**36**)

$C_{15}H_{24}O_2$; 236.35
m.p. 113°; $[\alpha]_D^{30}$ −24.9° (c 1.9)
Spectra: IR, ^{1}H NMR
^{1}**H NMR:** τ 9.32, 9.19, 9.13 &
9.02 (6H, gem-methyls), 8.35
(3H, vinylic methyl), 7.39 (2H, d,
J = 4 c/s, epoxide proton), 6.38
(1H, t, −C$\underline{H}_2$OH), 5.65 (1H,
OH), 4.55 (1H, =CH).
Derivatives: (i) Khusinodiol, m.p.
130°; $[\alpha]$ +22.7° (c 3.12) (ii)
Hydroxy acetate, m.p. 143°, $[\alpha]_D$
+18.0° (c 3.2).

Seshadri *et al.*, 1967

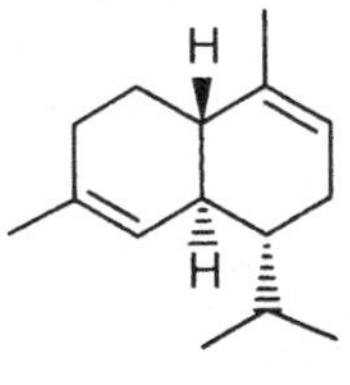

(+)-α-Amorphene (**37**)

$C_{15}H_{24}$; 204.35
$[\alpha]_D$ +120°
Spectra: IR, ^{1}H NMR
^{1}**H NMR:** δ (TMS, CDCl$_3$) 0.92
and 0.95 (3H each, d, 2 x CH$_3$),
1.63 (3H, vinyl-CH$_3$), 5.09 (1H,
vinyl-H), 5.34 1H, vinyl-H).

Andersen, 1970b

Table 4.5 (cont'd)

Cadinane Group		Ref.
α-Calacorene (**38**)	C$_{15}$H$_{20}$; 200.32 **Spectra:** UV, IR **Derivative:** on gas chromatographic dehydrogenation yielded cadalene.	Mizrahi and Nigam, 1969
Khusinodiol (**39**)	C$_{15}$H$_{26}$O$_2$; 238.37 m.p. 130°; [α] +22.7° (c 3.12)	Kalsi *et al.*, 1979a; Seshadri *et al.*, 1967
(+)-α-Cadinol (**40**)	C$_{15}$H$_{26}$O; 222.37 m.p. 75°; [α]$_D^{30}$ +49° **Spectra:** IR, ^{1}H NMR ^{1}**H NMR:** δ 0.77 and 0.89 (3H each, d, J = 7Hz, gem-methyls), 1.12 (3H, s, C(OH)C$\underline{H}_3$), 1.80 (3H, s, W$_{1/2}$ = 5Hz, CH$_3$), 5.49 (1H, s, W$_{1/2}$ = 6 Hz, =CH), 1.70 (1H, OH).	Kalsi *et al.*, 1979a
Cadina-4α,10β-diol (**41**)	C$_{15}$H$_{28}$O$_3$; 256.38 m.p. 117°; [α]$_D^{32}$ +60° **Spectra:** IR, ^{1}H NMR ^{1}**H NMR:** δ 0.75 and 0.87 (3H each, d, J = 7 Hz, gem-methyls), 1.02 and 1.21 (6H, s, 2 x CH$_3$)	Kalsi *et al.*, 1979a

Table 4.5 (cont'd)

Cadinane Group		Ref.
 Vetidiol (**42**)	$C_{15}H_{24}O_2$; 236.35 m.p. 170°; $[\alpha]_D^{30}$ −140° **Spectra**: IR, ^{1}H NMR **^{1}H NMR**: δ 0.78 and 0.90 (3H each, d, J = 7 Hz, gem-methyls), 1.8 (3H, s, =C−CH$_3$), 4.15 (1H, br m, $W_{1/2}$ = 20 Hz, −C<u>H</u>OH), 4.5 (1H, s, $W_{1/2}$ = 6 Hz, −C<u>H</u>OH), 5.05 and 5.2 (1H each, s, C=CH$_2$), 5.6 (1H, br s, −C=CH). **Derivatives**: (i) Acetate, IR, ^{1}H NMR (ii) Epoxide, m.p. 124°, IR, ^{1}H NMR (iii) (+)-α-Cadinol, m.p. 78°, $[\alpha]_D^{25}$ +52° (iv) (-)-γ-Cadineene (v) Isokhusinol, IR (vi) Khusinol, m.p. 87°.	Kalsi and Talwar, 1987
 Veticadinol (**43**)	$C_{15}H_{26}O$; 222.37	Chaurdoglu and Delesemme, 1961
 Compound **44**	$C_{15}H_{24}$; 204.35	Surve *et al.*, 1961
 Compound **45**	$C_{15}H_{24}$; 204.35 b.p. 94°/2 mm; $[\alpha]_D^{25}$ +68.24°; n_D^{31} 1.5029; d_4^{30} 0.9230 **Spectra**: IR **Derivatives**: (i) Mono-epoxide, b.p. 115° (bath)/0.0025 mm, IR (ii) Tetrahydro derivative, b.p. 83–84°/0.3 mm; $[\alpha]_D^{23}$ +7.39; n_D^{23} 1.4850.	Bhattacharyya *et al.*, 1959

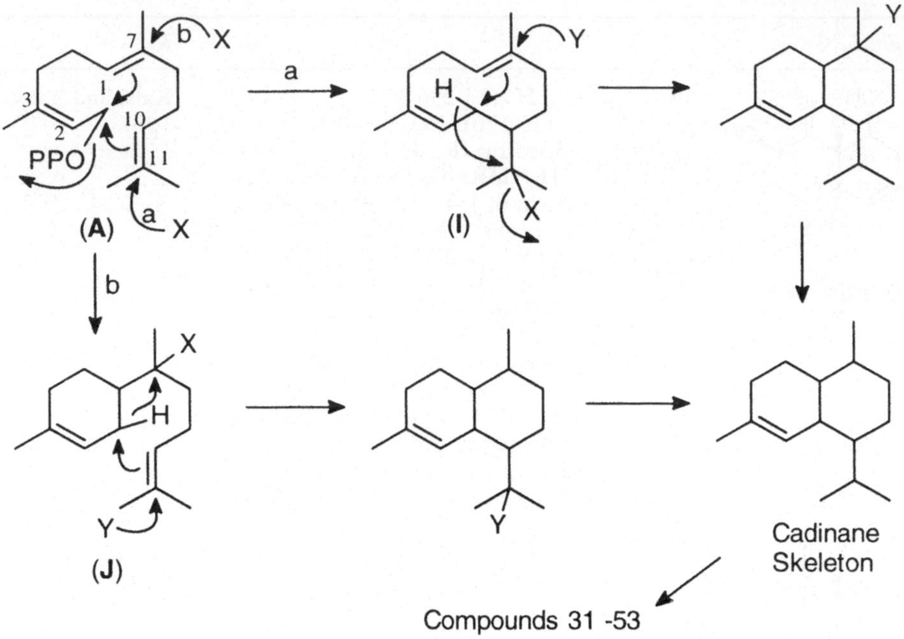

Figure 4.6 Cyclization of FPP to cadinane skeleton and compounds of the cadinane group. X & Y denotes enzymes.

by degradation. Compounds **44** and **45** were isolated (Bhattacharyya *et al.*, 1959; Surve *et al.*, 1961) from the dextrirotatory South Indian vetiver oil of the cultivated type.

Cyclization of farnesyl pyrophosphate (FPP, **A**) to the cadinane skeleton has been widely studied (Cordell, 1976; Akhila *et al.*, 1987). The mechanism has been demonstrated in Scheme 6 where X^- (enzyme) activated the cyclization process by attacking at C-11 of FPP (**A**) and the release of -OPP from C-1 (route a). On the other hand the cyclization could also be initiated by attack of Y^- (enzyme) at C-7 (route b). These two routes (a & b) would provide intermediates **I** and **J** respectively which can produce cadinane skeleton and all the compounds related to it through cyclization.

(v) Cadinane norsesquiterpene group

North Indian vetiver oil (*V. zizaniodes* Linn.) is unusually rich in biogenetically interesting antipodal terpenoids. A number of novel C_{14}-terpenoids possessing antipodal stereochemistry also occur in this oil. (+)-Khusitene (**47**) was isolated from the Punjab or Kuriala Ghat variety of North Indian vetiver oil (Raj *et al.*, 1971). The structure was given on the basis of chemical degradation and spectral evidences. Catalytic hydrogenation confirmed the presence of two double bonds one of which was found to be methylenic and the other a trisubstituted double bond, as indicated by IR and NMR data. It was found to be an optical antipode of khusilene (**46**) previously isolated by Trivedi (1966).

The carbonyl fraction of the North Indian vetiver oil afforded a novel aldehyde, khusilal (49) as a major constituent (Kalsi *et al.*, 1964). Further processing of the same fraction led to the isolation of small amounts of a new C_{14}-methyl ketone, khusitone (48), in pure form (Trivedi *et al.*, 1964). A primary alcohol khusilol (50), now isolated, was identified (Karkhanis *et al.*, 1978) with the $LiAlH_4$-reduction product of 49. An acid (51) was isolated in the form of its ester in pure form from the Punjab variety of Indian vetiver oil (Kalsi and Bhatia, 1969). This was identical with the corresponding aldehyde (49). The alcohol fraction of the oil, after removal of 31, 33 and 49, afforded a new C_{14}-ketoalcohol, khusitoneol (52) which was found to be active as a juvenile harmone against the mustard aphid (*Lipaphis erysmi*) resulting in the production of supernumerary instar nymphs, nymphal adult intermediates and adultoids (Kalsi *et al.*, 1985a).

Norkhusinol oxide (53), a new C_{14}-terpenoid, was isolated from the alcoholic fraction of North Indian vetiver oil (Kalsi *et al.*, 1985b). Its stereostructure was assigned on the basis of chemical correlation coupled with spectral data. The stereostructure was further confirmed by comparison of its plant growth activity causing root initiation in the hypocotyl cuttings of *Phaseolus aureus* with that displayed by khusinoloxide of known stereostructure. This was the first report in which biological evaluation was used as a tool to confirm stereostructure of a naturally occurring terpenoid.

C. Tricyclic sesquiterpenes

(i) Cedrane group

α-Cedrene (54), its oxygenated derivatives (55 & 56), (-)-α-funebrene (57), and its oxygenated derivatives (58–60) were isolated from vetiver oil by Kaiser and Naegeli (1972). The stereochemistry of these compounds was established on the basis of spectral and degradation results further supported by X-ray analysis (Paknikar *et al.*, 1975).

The cyclization of this group of compounds is again initiated by the attack of $\Delta^{6,7}$ on C-1 followed by attack of $\Delta^{10,11}$ on C-6, a 1,2 hydrogen shift to C-7 and release of nucleophile (enzyme) from C-7 to form intermediate K as shown in Scheme 7. This further cylizes to the cedrane group of compounds and acoradienes.

(ii) Zizaane and prezizaane group

Tricylovetivene or (+)-zizaene (61) was first isolated from vetiver oil in 1968 by Sakuma and Yoshikoshi (1968) and the structure was established by IR, ^1H NMR data and degradation analysis. Later on, it was reported in the Moosanagar variety of North Indian vetiver oil (Kirtany and Paknikar, 1971; Ganguli *et al.*, 1978a) and in vetiver oil of Javanese origin (Hanayama *et al.*, 1973). The stereo-selective total synthesis of 61 was given by Kido *et al.* (1969b) and Coates *et al.* (1972). (+)-prezizaene 62 was isolated for the first time by Andersen from the Reunion and Haiti varieties of the oil (Andersen and Falcon, 1971).

Ganguli *et al.* (1978a) isolated allokhusiol (63) along with 61 and 62 from the neutral fraction of the Moosanagar variety of vetiver oil; the structures were assigned on the basis of IR, ^1H NMR data and degradation analysis.

Khusiol (64), a new tricyclic, saturated, crystalline antipodal terpenic alcohol, has been isolated from the vetiver oil of the Moosanagar area (Ganguli *et al.*, 1978b). The structure was determined on the basis of chemical and spectral evidence. Khusiol on

Table 4.6

Cadinane norsesquiterpene Group		Ref.

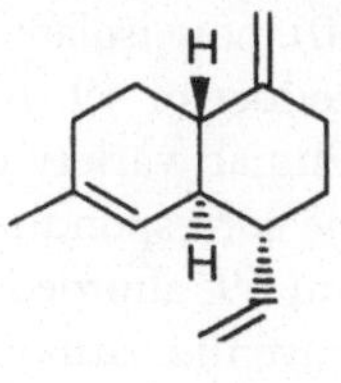

(+)-Khusilene (**46**)

$C_{14}H_{24}$; 188.31

Trivedi, 1966

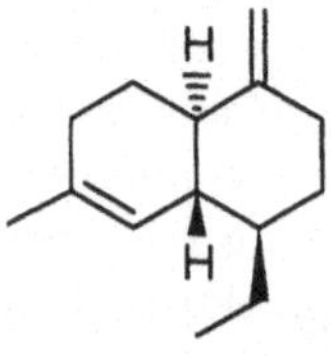

(+)-Khusitene (**47**)

$C_{14}H_{22}$; 190.32
b.p. 115° (bath)/2 mm; $[\alpha]_D^{30}$ +124°
(c 3.0)
Spectra: IR, ^{1}H NMR
^{1}**H NMR**: δ 0.9 (3H, t, $-CH_2-CH_3$), 1.66
(3H, s, allylic methyl), 4.35 and 4.45 (1H
each, d, J = 1.5 Hz, >C=CH_2), 5.5 (1H, br
q, J = 1.5 Hz, vinyl-H).
Derivatives: (i) Di- and tetrahydro-(+)-
khusitene, IR (ii) 4-Ethyl-1,
6-dimethylnaphthalene on Se
dehydrogenation, m.p. 135°
(iii) 1,2,5-Trimethylnaphthalene, IR.

Raj *et al.*, 1971

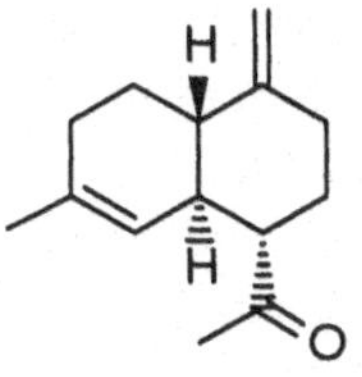

Khusitone (**48**)

$C_{14}H_{20}O$; 204.31
b.p. 118° (bath)/0.35 mm; $[\alpha]_D^{30}$
−134.8°; n_D^{29} 1.5060
Spectra: IR, ^{1}H NMR
^{1}**H NMR**: δ 1.65 (3H, s, allylic methyl), 2.1
(3H, s, >$COCH_3$), 4.6 and 4.7 (1H each, d,
>C=CH_2), 5.1 (1H, s, −CH=C<).
Derivatives: (i) Semicarbazone, m.p.
183–184° (ii) Khusitol, $[\alpha]_D^{27}$ −100°, IR
(iii) Dihydrokhusitone, b.p. 101° (bath)/0.6
mm, $[\alpha]_D^{26}$ +44.3° (c 0.88), n_D^{26} 1.4930
(iv) Tetrahydrokhusitone, b.p. 98° (bath)/
0.25 mm, n_D^{31} 1.4800 (v) Tertiary alcohol
with methyl lithium, m.p. 109–110°, IR.

Trivedi *et al.*, 1964

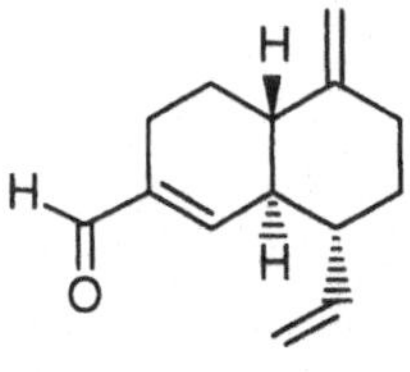

Khusilal (**49**)

$C_{14}H_{18}O$; 202.29
b.p. 115° (bath)/0.3 mm; $[\alpha]_D^{25}$ −26°;
n_D^{25} −261° (c 9.7)
Spectra: IR, UV, ^{1}H NMR
^{1}**H NMR**: δ 4.65 and 4.75 (1H each, d,
>C=CH_2), 5.1–5.5 (1H, m, vinyl-H), 6.62
(1H, s, −C$\underline{H}$=C−CHO), 9.4 (1H, s, −C$\underline{H}$O).
Derivatives: (i) 2,4-Dinitrophenylhydrazone,
m.p. 214°, UV (ii) Oxime, m.p. 101–102°,
UV (iii) Epoxykhusilal, b.p. 130° (bath)/0.8
mm, IR (iv) Khusilol, m.p. 74°, $[\alpha]_D^{24}$
−158°, IR, ^{1}H NMR (v) Dihydrokhusilol,
m.p. 72°, $[\alpha]_D^{24}$ −30° (vi) Dihydrokhusilal,
b.p. 110° (bath)/0.5 mm, IR (vii)
Hexahydrokhusilol, b.p. 130° (bath)/0.9 mm,
$[\alpha]_D^{27}$ +44° (c 1.4), n_D^{27} 1.4965 (viii)
Hexahydrokhusileene, b.p. 103° (bath)/3
mm, $[\alpha]_D^{30}$ +44° (c 3), n_D^{30} 1.4761, IR,
^{1}H NMR.

Kalsi *et al.*, 1964

Table 4.6 (cont'd)

Cadinane norsesquiterpene Group		Ref.
Khusilol (**50**)	$C_{14}H_{20}O$; 204.31 m.p. 74°, $[\alpha]_D^{24}$ −158° **Spectra:** IR, ^{1}H NMR ^{1}**H NMR:** δ 1.64 (1H, s, =C–CH$_2$OH), 3.98 (2H, s, =C–CH$_2$OH), 4.52 and 4.6 (1H each, d, >C=CH$_2$), 5.0–5.5 (1H, m, –CH=CH$_2$), 5.68 (1H, s, –CH=C–CH$_2$OH).	Kalsi *et al.*, 1964; Karkhanis *et al.*, 1978
Acid (**51**)	$C_{14}H_{18}O_2$; 218.29 m.p. 124°; $[\alpha]_D^{24}$ −198° (c 2) **Spectra:** IR, UV	Kalsi *et al.*, 1964; Kalsi and Bhatia, 1969
Khusitoneol (**52**)	$C_{14}H_{20}O_2$; 220.31 Colourless liquid $[\alpha]_D^{30}$ −140° **Spectra:** UV, IR, ^{1}H NMR ^{1}**H NMR:** δ 1.72 (3H, s, vinylic methyl), 2.1 (3H, s, –COCH$_3$), 1.9 (1H, exchangeable with D$_2$O, –OH), 4.29 (1H, –CHOH), 4.55 (2H, >C=CH$_2$), 5.50 (1H, >C=CH). **Derivatives:** (i) Khusitone (ii) Tetrahydrokhusitoneol	Kalsi *et al.*, 1985a
Norkhusinol oxide (**53**)	$C_{14}H_{22}O_2$; 222.32 m.p. 76°; $[\alpha]_D^{20}$ −120° **Spectra:** UV, IR, ^{1}H NMR ^{1}**H NMR:** δ 0.95 (3H, t, J = 7 Hz, –CH$_2$–CH$_3$), 1.62 (3H, s, >C=C–CH$_3$), [2.25 (1H, d, J = 4 Hz) and 2.8 (1H, dd, J = 2, 4 Hz) protons on epoxide], 3.7 (1H, br m, W$_{1/2}$ = 20 Hz, –CHOH) 5.5 1H, br s, >C=CH). **Derivative:** (i) 1,6-Dimethyl-4-ethyl naphthalene (ii) Diol m.p. 87°.	Kalsi *et al.*, 1985b

treatment with CrO$_3$ in pyridine gave a six-membered ketone, khusione, which showed a broad singlet at δ 2.12 integrating for two protons in its NMR spectrum. This broad singlet was due to lack of any vicinal coupling for the C$_{10}$-methylene protons attached to the carbonyl group, and therefore suggested a –C–CH$_2$–C=O group in the molecule. The rearrangement of **64** to **62** could be envisaged by the removal of the C$_{11}$-hydroxyl of **64** with the migration of the C$_8$-methylene and subsequent deprotonation. Thus this rearrangement proves the structure and absolute configuration of **64** except for the stereochemistry of the OH-group. R-configuration at the hydroxyl position was indicated by the difference of molecular rotation between the *p*-nitrobenzoate derivative and the parent alcohol, which was −96.5°.

Table 4.7

Cedrene Group		Ref.
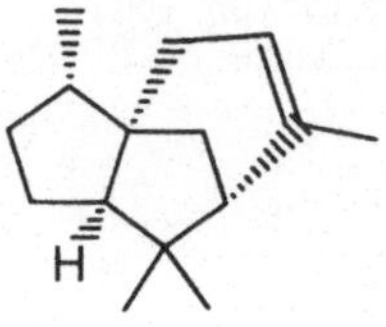 Cedrene (**54**)	$C_{15}H_{24}$; 204.35 $[\alpha]_D$ +81° (CHCl$_3$, c 0.5)	Kaiser and Naegeli, 1972
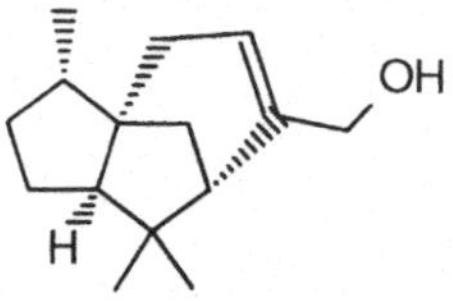 Cedrenol (**55**)	$C_{15}H_{24}O$; 220.35 **Spectra**: IR, MS, ^{1}H NMR ^{1}**H NMR**: δ (TMS, CDCl$_3$) 0.95 and 0.97 (3H each, s, 2 x CH$_3$), 0.86 (3H, d, J = 6.5 Hz, −CH$_3$), 3.97 (2H, s, −C<u>H</u>$_2$OH), 5.50 (1H, m, >C=CH).	Kaiser and Naegeli, 1972
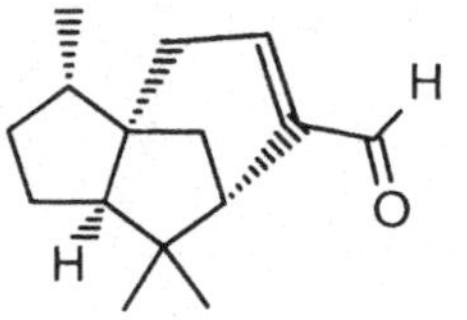 Cedrenal (**56**)	$C_{15}H_{22}O$; 218.33 **Spectra**: IR, MS, ^{1}H NMR ^{1}**H NMR**: δ (TMS, CDCl$_3$) 0.88 and 1.06 (3H each, s, 2 x CH$_3$), 0.95 (3H, d, J = 6.5 Hz, −CH$_3$), 2.39 (2H, m as pseudo dd, C=C− CH$_2$−), 2.69 (1H, br d, J = 4 Hz, >C=C(CHO)−C<u>H</u>−), 6.72 (1H, pseudo t, J = 3.5 Hz, >C=CH−), 9.53 (1H, s, −CHO).	Kaiser and Naegeli, 1972
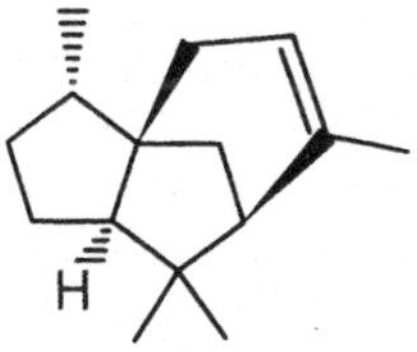 (-)-α-Funebrene (**57**)	$C_{15}H_{24}$; 204.35 $[\alpha]_D$ −102° (CHCl$_3$, c 1.176) **Spectra**: IR, MS, ^{1}H NMR ^{1}**H NMR** : δ (TMS, CDCl$_3$) 0.85 and 1.06 (3H each, s, 2 x CH$_3$), 0.87 (3H, d, J = 6.5 Hz, −CH$_3$), 1.63 (3H, m, C=C−CH$_3$), 5.12 (1H, m, >C=CH−).	Kaiser and Naegeli, 1972; Paknikar *et al.*, 1975
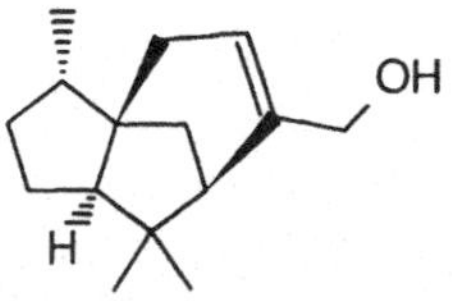 Funebrenol (**58**)	$C_{15}H_{24}O$; 220.35 **Spectra**: IR, MS, ^{1}H NMR ^{1}**H NMR**: δ (TMS, CDCl$_3$) 0.835 and 1.095 (3H, s, 2 x CH$_3$), 0.89 (3H, d, J = 7 Hz, −CH$_3$), 3.95 (2H, dd, J = 2 Hz & 4 Hz, >C=C−CH$_2$−), 5.41 (1H, >C=CH−).	Kaiser and Naegeli, 1972; Paknikar *et al.*, 1975

Table 4.7 (cont'd)

Cedrene Group		Ref.
 Funebrenal (**59**)	$C_{15}H_{22}O$; 218.33 **Spectra:** IR, MS, ^{1}H NMR ^{1}**H NMR:** δ (TMS, CDCl$_3$) 0.65 and 1.12 (3H each, s, 2 x CH$_3$), 0.91 (3H, d, J = 6.5 Hz, −CH$_3$), 2.29 and 2.48 and 3.06 (1H each, d, J = 4 Hz, >CH−C=C−CH$_2$−), 6.55 (1H, m, >C=CH−), 9.30 (1H, s, −CHO).	Kaiser and Naegeli, 1972; Paknikar *et al.*, 1975
 Funebrene acid (**60**)	$C_{15}H_{22}O_2$; 234.33 **Spectra:** IR, MS, ^{1}H NMR ^{1}**H NMR:** δ (TMS, CDCl$_3$) 0.75 and 1.12 (3H each, s, −CH$_3$), 0.91 (3H, d, J = 6.5Hz, −CH$_3$), 3.0 (1H, d, J = 4.5 Hz), 3.72 (3H, s), 6.72 (1H, m, vinylic proton).	Kaiser and Naegeli, 1972; Paknikar *et al.*, 1975

Khusimol (**65**) was first isolated from the high boiling fraction using column chromatography and its structure was given by IR and NMR (Umarani *et al.*, 1966). Later on, long-range coupling was demonstrated in its NMR (Neville and Nigam, 1969). **65** was named khusenol in one of the publications appearing in 1968 (Nigam *et al.*, 1968). However, IR and ^{1}H NMR data of khusenol suggested its structure as **65**. Khusimone (**66**) was isolated from vetiver oil (Umarani *et al.*, 1970) as a highly odoriferous ketone and its absolute chemistry suggested its structure as **66** by spectroscopic and chemical evidence. A total synthesis of **66** has been achieved in 16 steps from the ammonium salt of *l*-10-camphorsulphonic acid (Liu and Chan, 1979).

Khusenic acid (**67**) and isokhusenic acid (**68**) were isolated from vetiver oil (Nigam and Komae, 1967; Komae and Nigam, 1968; Hanayama *et al.*, 1973) and dehydrogenation results suggested that the two acids differ in the location of the double bond. **67** possesses a terminal methylene group which rearranges in the presence of mineral acid to a tetra-substituted double bond leading to the formation of **68**. The methyl esters of these acids showed the base peak of m/e 102 in mass spectra (Kido *et al.*, 1967), thus indicating that all these four compounds bear a gem-dimethyl group on C-2. Methyl khusenate (**70**) and methyl isokhusenate (**71**) also differed in the position of the double bond as in **67** and **68**. The structures were assigned on the basis of spectral analysis.

Epizizanoic acid (**69**) was isolated along with zizanoic (khusenic) acid (**67**) from the vetiver oil and its structure was confirmed on the basis of chemical degradation (Hanayam *et al.*, 1973). A complete degradation of **61**, **65**, **67** and **69** has been conducted to prove that all these sesquiterpenoids belong to the zizaane family.

Zizanol (**72**), isolated as a colourless oil, was a mono-unsaturated tricyclic secondary alcohol displaying a positive RD curve, $[\alpha]_D$ +28° (Andersen, 1970a). In C_6D_6 the CH$_3$ region of the ^{1}H NMR was clearer than in CDCl$_3$, both lines of the doublet CH$_3$ (δ 0.93) appearing upfield from the six proton singlet due to the gem-dimethyl (δ

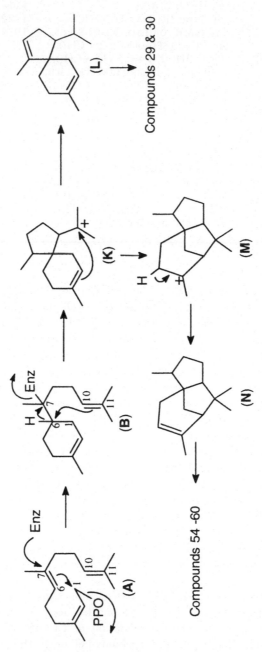

Figure 4.7 Suggested biosynthetic pathway to the acoradienes and the cedrane group of compounds.

Table 4.8

Zizaane & prezizaane Group		Ref.
 Tricyclovetivene or (+)-Zizaene (61)	$C_{15}H_{24}$; 204.35 b.p. 130–35° (bath)/2.0 mm; $[\alpha]_D$ +43° (c 1.2) **Spectra:** IR, MS, ^{1}H NMR ^{1}**H NMR:** δ 0.99 (3H, d, J = 7 Hz, −CH$_3$), 1.03 and 1.06 (3H each, s, 2 x CH$_3$), 4.68 and 4.53 (1H each, t, J = 1.5 Hz, >C=CH$_2$). **Synthesis** **Derivatives:** (i) OsO$_4$ oxidation to Glycol, m.p. 84–84.5°, UV, ^{1}H NMR (ii) nor-Ketone on ozonization.	Coates *et al.*, 1972; Ganguli *et al.*, 1978a; Hanayama *et al.*, 1973; Kido *et al.*, 1969b; Kirtany and Paknikar, 1971; Sakuma and Yoshikoshi, 1968
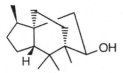 **Prezizaene (62)**	$C_{15}H_{24}$; 204.35 b.p. 120–125° (bath)/1.5 mm; $[\alpha]_D$ +53° (c 1.6) **Spectra:** IR, ^{1}H NMR, MS ^{1}**H NMR:** δ 0.85 (3H, d, J = 7 Hz, −CH$_3$), 1.05 Nd 1.08 (3H each, s, 2 x CH$_3$), 2.8 (1 H, br t), 4.62 (2H, br s, >C=CH$_2$).	Andersen and Falcone, 1971; Ganguli *et al.*, 1978a
 Khusiol (63)	$C_{15}H_{26}O$; 222.37 m.p. 81°; $[\alpha]_D$ −66° (c 1.4) **Spectra:** IR, MS, ^{1}H NMR ^{1}**H NMR:** δ 0.73, 0.86 and 0.89 (3H each, s, 3 x CH$_3$), 0.81 (3H, d, J = 6 Hz, sec. Methyl), 1.28 (1H, br s, disappeared on D$_2$O exchange, −OH), 3.86 (1H, octet, J = 2 Hz, 6 Hz, 9 Hz, >CHOH). **Derivatives:** (i) Khusione, b.p. 115–118° (bath)/0.4 mm; $[\alpha]_D$ −42° (c 1.8), IR, MS, ^{1}H NMR (ii) Khusiane, $[\alpha]_D$ +33.5° (c 1.61), IR, MS (iii) Khusiene, IR, MS	Ganguli *et al.*, 1978b
 Allokhusiol (64)	$C_{15}H_{26}O$; 222.37 b.p. 130–135° (bath)/1 mm; $[\alpha]_D$ +45° (c 1.22) **Spectra:** IR, MS, ^{1}H NMR ^{1}**H NMR:** δ 0.88 (6H, s, 2 x CH$_3$), 0.9 (3H, d, J = 6 Hz, Sec. Methyl), 1.1 (3H, s, CH$_3$). **Derivatives:** (i) (+)-Prezizaene (ii) (+)-Zizaene **Biosynthesis**	Akhila *et al.*, 1987; Ganguli *et al.*, 1978a

Table 4.8 (cont'd)

Zizaane & prezizaane Group		Ref.

Khusimol, Khusenol or cyclovetivenol (65)

$C_{15}H_{24}O$; 220.35
b.p. 140–145°/0.07 mm; $[\alpha]_D^{27}$ +24.56° ($CHCl_3$, c 4.2); n_D^{24} 1.5183
Spectra: IR, ^1H NMR
^1H NMR: δ 1.01 (6H, s, gem-methyls), 2.5 (1H, s, −OH), 3.38 (2H, −C$\underline{H}_2$OH), 4.38 and 4.51 (1H each, d, >C=CH_2).
Derivatives: (i) Unsaturated acid, $C_{15}H_{22}O_2$, IR (ii) Saturated alcohol, $C_{15}H_{26}O$, b.p. 121–128°/0.3 mm, n_D^{24} 1.507, $[\alpha]_D^{27}$ +33.44° (c 4.1) (iii) Unsaturated hydrocarbon, $C_{15}H_{24}$, b.p. 165–170°/3.5 mm, n_D^{25} 1.497, $[\alpha]_D^{25}$ +45.64°, IR, ^1H NMR
Biosynthesis

Akhila *et al.*, 1987; Neville and Nigam, 1969; Nigam *et al.*, 1968; Umarani *et al.*, 1966, 1969

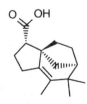

Khusimone (66)

$C_{14}H_{20}O$; 204.31
Synthesis

Liu and Chan, 1979; Umarani *et al.*, 1970

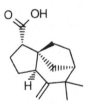

Khusenic or Zizanoic acid (67)

$C_{15}H_{22}O_2$; 234.33
$b_{0.5}$ 158°; $[\alpha]_D$ +17.2° (c 4.39); n_D^{26} 1.5198
Spectra: IR, ^1H NMR
^1H NMR: δ (TMS, CCl_4) 1.05 and 1.08 (3H each, s, 2 x CH_3), 2.62 (1H, q, >C$\underline{H}$COOH), 4.53 and 4.70 (1H each, t, >C=CH_2), 11.08 (1H, s, −COOH).
Derivatives: (i) Cyclohexylamine salt, m.p. 146° (ii) Methyl ester, $C_1H_{24}O_2$, b.p. 120–121°/1.8 mm, $[\alpha]_D^{20}$ +39.7° (MeOH, c 1.2), IR, ^1H NMR.
Synthesis

Hanayama *et al.*, 1973; Kido *et al.*, 1967; Komae and Nigam, 1968; MacSweeney and Ramage, 1971; Nigam & Komae, 1967

Isokhusenic acid (68)

$C_{15}H_{22}O_2$; 234.33
m.p. 80–81°; $[\alpha]_D$ +27.9° (c 2.78)
Spectra: IR, ^1H NMR
Synthesis

Kido *et al.*, 1972; Komae and Nigam, 1968; MacSweeney and Ramage, 1971; Nigam and Komae, 1967

Table 4.8 (cont'd)

Zizaane & prezizaane Group		*Ref.*

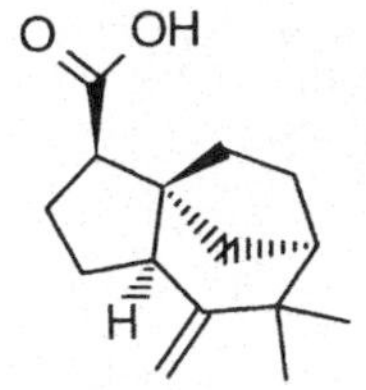

Epizizanoic acid (**69**)

C$_{15}$H$_{22}$O$_2$; 234.33
m.p. 109–110.5°
Spectra: IR, ^{1}H NMR
^{1}**H NMR**: δ (TMS, CCl$_4$) 1.05
and 1.09 (3H each, s, 2 x CH$_3$),
4.57 and 4.77 (1H each, t,
J = 1.5 Hz)
Synthesis

Hanayama *et al.*, 1968;
Hanayama *et al.*, 1973;
Kido *et al.*, 1972

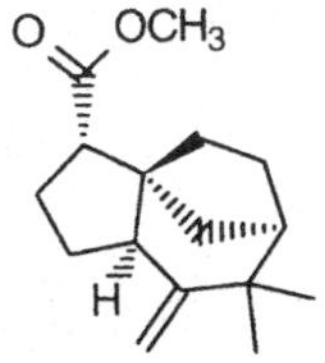

Methyl khusenate (**70**)

C$_{16}$H$_{24}$O$_2$; 248.36
b.p. 120–121°/1.8 mm; [α]$_D$
+39.7° (MeOH, c 1.2)
Spectra: IR, ^{1}H NMR, MS
^{1}H NMR: δ 1.20 and 1.23 (3H
each, s, 2 x CH$_3$), 2.58 (1H, q,
J = 4 and 7 Hz), 3.60 (3H, s,
−COOCH$_3$), 4.5 and 4.7 (1H
each, t, J = 1.5 Hz, >C=CH$_2$).
Derivatives: (i) Zizanoic acid
(ii) Methyl dihydrozizanoate, b.p.
122–125°/1.8 mm, IR
(iii) Khusimol

Hanayama *et al.*, 1973;
Neville and Nigam,
1969; Nigam and
Komae, 1967

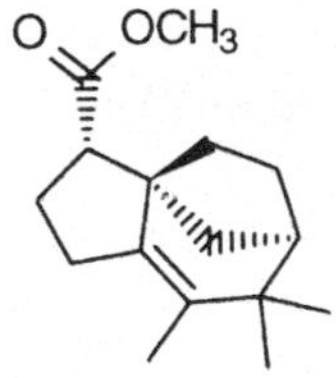

Methyl isokhusenate
(**71**)

C$_{16}$H$_{24}$O$_2$; 248.36
m.p. 51°; [α]$_D$ +46.8° (c 1.64)
Spectra: IR, ^{1}H NMR
^{1}**H NMR**: δ 1.00 (6H, s, 2 x
CH$_3$), 1.46 (3H, t, J = 1.0 Hz,
CH$_3$), 3.64 (3H, s, −COOCH$_3$).
Derivatives: (i) Isokhusenic acid,
m.p. 80–81°; [α]$_D$ +27.9°
(c 2.78), IR, NMR

Neville and Nigam,
1969; Nigam and
Komae, 1967

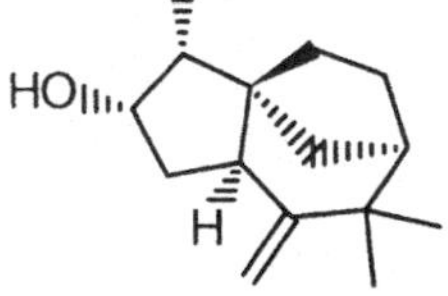

Zizanol (**72**)

C$_{15}$H$_{24}$O; 220.35
[α]$_D^{18}$ +10.4° (ref. 40), [α]$_D$
+28° (ref. 8)
Spectra: IR, ^{1}H NMR, MS
^{1}**H NMR**: δ (TMS, CDCl$_3$) 1.04
(3H, d, J = 7.0, CH$_3$), 1.06 (6H,
s, gem-methyls), 2.52 (1H, m,
>C=C−CH<), 3.83 (1H, dt,
J$_d$ = 3.1 and J$_t$ = 7.1 Hz,
>C$\underline{H}$OH), 4.56 and 4.73 (1H
each, t, J = 1.7Hz, >C=CH$_2$).
Derivative: (i) Zizanone, m.p.
51°, IR, ^{1}H NMR (ii) Dihydro
derivative, ^{1}H NMR (iii) Triol,
m.p. 94–95°.

Andersen, 1970a

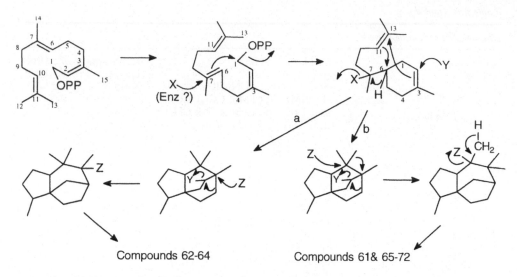

Figure 4.8 Mechanism for the biogenesis of zizaane and prezizaane compounds.

1.045). The zizaene skeleton, suggested by the >C–C=CH$_2$ grouping and its behaviour in NMDR experiments, was confirmed via reduction (LiAlH$_4$/THF) of the tosylate which afforded a single hydrocarbon identified as zizaene by GLC comparison.

The biosynthesis of two representative compounds khusimol (65) of the zizaane group and allokhusiol (64) of the prezizaane group has been studied in detail by Akhila *et al.* (1987) using radioactive precursors. It has been proposed for the biosynthesis of 65 and 64 (Figure 4.8) that **A** would cyclize to **B** with X (enzyme or its biogenetic equivalent) attacking at C-7, double bond attacking at C-1 and -OPP being lost. **B** is further cyclized to **O** with Y attacking at C-2, $\Delta^{2,3}$ attacking at C-11, hydride shift from C-6 to C-7 and release of nucleophile (X) from C-7. **O** is the common precursor for zizaane and prezizaane compounds.

D. Tetracyclic sesquiterpenes

(i) Cyclocopacamphene group

Typical vetiver oils have been shown to contain cyclocopacamphene (73) and related oxygenated sesquiterpenes (Kido *et al.*, 1969a; Andersen, 1970a, 1970b). They were the first sesquiterpene constituents of typical vetiver oils to bear a 7α-isopropyl group. Cyclocopacamphenol and epicyclocopacamphenol (74 & 75) were isolated from vetiver oil as an inseparable mixture (Andersen, 1970a; Homma *et al.*, 1970). The ^{1}H NMR data given in Table 4.9 was indicated by the whole mixture. Jones oxidation of the above mixture provided a mixture of carboxylic acids, from which two carboxylic acids, identified as cyclocopacamphenic and epicyclocopacamphenic acids (76 & 77), were separated by fractional crystallization. These carboxylic acids (76 & 77) were also isolated (Kido *et al.*, 1969a) from vetiver oil in pure form by mild alkaline hydrolysis of methyl esters separated by preparative GLPC of the whole ester mixture. Epimeric relationship of the carboxyl groups in these acids was unambiguously confirmed by

Table 4.9

Cyclocopacamphene Group		Ref.
Cyclocopacamphene (73)	$C_{15}H_{24}$; 204.35 $[\alpha]_D^{19.5}$ +35.0° (CHCl$_3$, c 1.6) Spectra: IR, ^{1}H NMR ^{1}H NMR: δ (TMS, CCl$_4$) 0.62, 0.67, 0.75 (3H, s), 0.88 and 0.91 (3H each, d, J = 6.5 Hz), 1.02 (3H, s).	Andersen, 1970a, 1970b; Kido *et al.*, 1969a
Cyclocopacamphenol and Epicyclocopacamphenol (74 & 75)	$C_{15}H_{24}O$; 220.35 Spectra: IR, ^{1}H NMR ^{1}H NMR: δ (TMS, CDCl$_3$) 0.75 (3H, s, CH$_3$), 1.0 (3H, d, CH$_3$), 1.02 (3H, s, CH$_3$), 3.42 and 3.78 (2H, ddd, J = 3.0 Hz, 5.5 Hz and 10.5 Hz, −CH$_2$OH) **Derivatives**: Cyclocopacamphenic and epicyclocopacamphenic acids.	Andersen, 1970a; Homma *et al.*, 1970
Cyclocopacamphenic acids (76)	$C_{15}H_{22}O_2$; 234.33 m.p. 151.5–152.5°; $[\alpha]_D^{19}$ −14.7° (CHCl$_3$, c 1.25) Spectra: IR, ^{1}H NMR ^{1}H NMR: δ (TMS, CCl$_4$) 0.77 (3H, s, CH$_3$), 1.03 (3H, s, CH$_3$), 1.19 (3H, d, J = 7 Hz, CH$_3$).	Kido *et al.*, 1969a
Epicyclocopacamphenic acid (77)	$C_{15}H_{22}O_2$; 234.33 m.p. 168–168.5°; $[\alpha]_D^{20.5}$ +78.3° (CHCl$_3$, c 3.6) Spectra: IR, ^{1}H NMR ^{1}H NMR: δ (TMS, CDCl$_3$) 0.77 (3H, s, CH$_3$), 1.02 (3H, s, CH$_3$), 1.17 (3H, d, J = 6 Hz, CH$_3$).	Kido *et al.*, 1969a

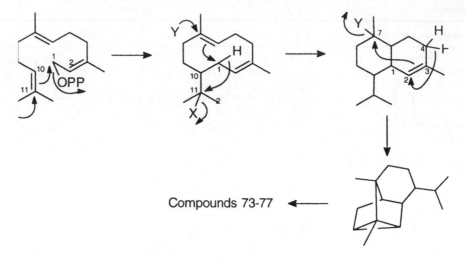

Figure 4.9 Cyclization of FPP to cyclocopacamphane compounds *via* the cadinane
skeleton.

derivation to the same parent hydrocarbon from both acids. However, the absolute
configuration of the carboxyl group in these two acids could not be assigned. Spectral
and chemical evidence indicated that these acids are saturated tetracyclic compounds
possessing one secondary and two tertiary methyl groups; furthermore, the absorption
bands at near 3050, 860 and 845 cm^{-1} in the IR spectrum and at around δ 0.5–0.9
ppm (m) in the NMR spectrum in all of the derivatives suggested the presence of a
tricyclene carbon framework.

A hypothetical biosynthetic scheme has been projected in Figure 4.9; the initial
steps of cyclization are the same as in the case of other skeletons. The unusual 1,3 –
hydrogen shift forms the cyclopropane ring of the cyclocopacamphane skeleton.

References

Agarwal, R.C., Karkhanis, D.W., Audichya, T.D., Trivedi, G.K. and Bhattacharya, S.C. (1985)
O-Methylation of Khusinol and its Derivatives and a One-step Introduction of Tertiary
Methoxy Function in Khusinol. *Ind. J. Chem.*, 24**B**(2), 159.

Akhila, A., Sharma, P.K. and Thakur, R.S. (1987) Biosynthesis of Khusimol and Allokhusimol
in *Vetiveria zizanioides*. *Fitoterapia*, **58**(4), 243.

Akhila, A., Nigam, M.C. and Virmani, O.P. (1981) Vetiver Oil – A Review of Chemistry.
Curr. Res. On Medicinal and Aromatic Plants, 3(3), 195.

Andersen, N.H. and Falcone, M.S. (1971) (+)-Prezizaene and the Biogenesis of Zizaene. *Chem.
and Ind.*, 62.

Andersen, N.H., Falcone, M.S. and Syrdal, D.D. (1970d) Structure of Vetivenenes and
Vetispirenes. *Tetrahedron Letters*, 1759.

Andersen, N.H. (1970c) Biogenetic Implications of the Antipodal Sesquiterpenes of Vetiver
Oil. *Phytochemistry*, **9**, 145.

Andersen, N.H. (1970a) The Structure of Zizanol and Vetiselinenol. *Tetrahedron Letters*, 1755.

Andersen, N.H. (1970b) On the Occurrence of Levojunenol and Zizanene [(+)-α-Amorphene].
Tetrahedron Letters, 4651.

Bhattacharyya, S.C., Chakravarti, K.K. and Surve, K.L. (1959) A New Cadelenic Hydrocarbon. *Chem. Ind.*, 16, 1352.

Bhattacharyya, S.C., Rao, A.S. and Shaligram, A.M. (1960) The Absolute Configurations of Junenol and *Laevo*-junenol. *Chem. Ind.*, 17, 469.

Bhatwadekar, S.V., Pednekar, P.R., Chakravarti, K.K. and Paknikar, S.K. (1982) A Survey of Sesquiterpenoids of Vetiver Oil in C.K. Atal and B.M. Kapoor Cultivation and Utilization of Aromatic Plants. PID, New Delhi, p. 412.

Buchi, G.H. (1978) Total Synthesis of Spirovetivanes and Khusimone. *Perfumer & Flavorist*, 3, 2.

Buchi, G.H., Berthet, D., Decorzant, R., Grieder, A. and Hauser, A. (1976) Spirovetivanes from Fulvenes. *J. Org. Chem.*, 41, 3208.

Buddhsukh, D. and Magnus, P. (1975) Synthesis of (+)-Hinesol and 10-*epi*-(+)-Hinesol. *J. C. S., Chem. Commun.*, 952.

Caine, D., Boucugnani, A.A., Chao, S.T., Dawson, J.B. and Ingwalson, P.F. (1976) Stereospecific Synthesis of 6,c-10-Dimethyl(r-5-C')spiro[4.5]dec-6-en-2-one and Its Conversion into (±)-α-Vetispirene. *J. Org. Chem.*, 41, 1539.

Chiurdoglu, G. and Delesemme, A. (1961) Sesquiterpenes. V. Structure of Bicyclovetivenol and Veticadinol, tertiary alcohols of the oil of Vetiver from the Congo. *Bull des Societes Chimiques Belges,* 70(1–2), 5; *Perfum. Essent. Oil Rec.*, 52(7), 438.

Coates, R.M., Farney, R.F., Johnson, S.M. and Poul, I.C. (1969) Crystal Structure of Khusimol *p*-Bromobenzoate. *J. C. S. Chem. Commun.*, 999.

Coates, R.M. and Sowerby, R.L. (1972) Stereoselective Total Synthesis of (+)-Zizaene. *J. Am. Chem. Soc.*, 94, 5386.

Cordell, G.A. (1976) Biosynthesis of Sesquiterpenes. *Chem. Rev.*, 76, 425.

Dastur, K.P. (1974) A Stereoselective Approach to Eremophilane Sesquiterpenes. A Synthesis of (±)-Nootkatone and (±)-α-Vetivone. *J. Am. Chem. Soc.*, 96, 2605.

Dauben, W.G. and Hart, D.J. (1975) The Total Synthesis of Spirovetivanes. *J. Am. Chem. Soc.*, 97, 1622.

Deighton, M., Hughes, C.R. and Ramage, R. (1975) Stereospecific Synthesis of (-)-Agarospirol and (-)-β-Vetivone. *J. C. S., Chem. Commun.*, 662.

Dhingra, S.N., Dhingra, D.R. and Bhattacharyya, S.C. (1956) Structure of Khusol. *Perfum. Essent. Oil Rec.*, 47(10), 350.

Endo, K. and Mayo, P. de. (1967) α-Vetivone. *J. C. S., Chem. Commun.*, 89.

Fuehrer, H. (1970) Vetiver Oil and Vetiver Fragrances in Modern Perfumery. *Dragoco Rep.*, 17(10), 217.

Fuehrer, H. (1974) Vetiver Oils and Modern Wood and Vetiver Perfumes. *Fette, Seifn, Anstrichm*, 76, 40.

Ganguli, R.N., Trivedi, G.K. and Bhattacharyya, S.C. (1978a) Isolation of Allokhusiol and Biogenesis of Vetiver Tricyclics. *Ind. J. Chem.*, 16B, 20.

Ganguli, R.N., Trivedi, G.K. and Bhattacharyya, S.C. (1978b) Khusiol, a Biogenetically Significant Component from Vetiver Oil. *Ind. J. Chem.*, 16B, 23.

Garnero, J. (1972) Composition of Vetiver Essential Oils. *Riv. Ital. Essenze, Profumi Piante Off., Aromi, Saponi, Cosmet., Aerosol*, 54, 315.

Gill, H.S., Wadia, M.S., Bhatia, I.S. and Kalsi, P.S. (1969) Structure of Khusilal. *Chem. Ind.*, 1779.

Grove, J.F., Poppi, R. and Radley, M. (1968) The Biological Activity of Gibberellin Analogues: Zizanoic Acid. *Phytochem.*, 7, 2001.

Hanayama, N., Kido, F., Sakuma, R., Uda, H. and Yoshikoshi, A. (1968) Minor Acidic Constituents of Vetiver Oil. *Tetrahedron Letters*, 58, 6099.

Hanayama, N., Kido, F., Tanaka, R., Uda, H. and Yoshikoshi, A. (1973) Sesquiterpenoids of Vetiver Oil – I. The Structure of Zizanoic Acid and Related Constituents. *Tetrahedron*, 29, 945.

Homma, A., Kato, M., Wu, M.D. and Yoshikoshi, A. (1970) Minor Sesquiterpene Alcohols of Vetiver Oil. *Tetrahedron Letters*, 231.

Ishida, T., Nishimura, M., Hayashi, S., Matsura, T. and Araki, M. (1970) Identification of Valencene as a Sesquiterpene of Camphor Oil. *Chem. Ind.*, 312.

Jentsch, J. and Treibs, W. (1968) Constitution of Vetiver Oil Components, especially of the Tertiary Vetivenols. I. Isolation and Dehydrogenation of *tert.*-bicyclovetivenol and *tert.*-tricyclovetivenol. *Parfumi. Kosmet.*, 49(2), 29.

Jones, R.V.H. and Sutherland, M.D. (1968) Hedycaryol, the Precursor of Elemol. *J. C. S. Chem. Commun.*, 1229.

Kaiser, R. and Naegeli, P. (1972) Biogenetically Significant Components in Vetiver Oil. *Tetrahedron Letters*, No. 20, 2009.

Kalsi, P.S., Chakravarti, K.K. and Bhattacharyya, S.C. (1962) Terpenoids-XXXV. The Structure of Isobisabolene, a New Sesquiterpene Hydrocarbon from Vetiver Oil. *Tetrahedron*, 18, 1165.

Kalsi, P.S., Chakravarti, K.K. and Bhattacharyya, S.C. (1963) Terpenoids-XL. The Structure and Absolute Configuration of Khusol. *Tetrahedron*, 19, 1073.

Kalsi, P.S., Chakravarti, K.K. and Bhattacharyya, S.C. (1964) Terpenoids-LIII. Structure of Khusilal, a Novel Aldehyde from Vetiver Oil. *Tetrahedron*, 20, 2617.

Kalsi, P.S. and Bhatia, I.S. (1969) A New Acidic Component from the Indian Vetiver Oil. *Curr. Sci.*, 38, 363.

Kalsi, P.S., Kohli, J.C. and Wadia, M.S. (1972) Structure and Absolute Configuration of Epikhusinol, a New Sesquiterpene Alcohol from Vetiver Oil. *Ind. J. Chem.*, 10, 1127.

Kalsi, P.S., Arora, G.S. and Ghulati, R.S. (1979a) New Antipodal Sesquiterpene Alcohols from Vetiver Oil. *Phytochem.*, 18(7), 1223.

Kalsi, P.S., Arora, G.S. and Chhina, K. (1985a) Structure and Absolute Configuration of Khusitoneol, a New C_{14}-Ketoalcohol from Vetiver Oil. *Ind. J. Chem.*, 24B, 496.

Kalsi, P.S., Kaur, B. and Talwar, K.K. (1985b) Stereostructure of Norkhusinoloxide, a New Antipodal C_{14} Terpenoid from Vetiver Oil. Confirmation of Stereostructural Features by Biological Evaluation, a New Tool for Prediction of Stereostructure in Cadinanes. *Tetrahedron*, 41(16), 3387.

Kalsi, P.S. and Talwar, K.K. (1987) Stereostructure of Vetidiol, a New Antipodal Sesquiterpene diol from Vetiver Oil; A Novel Role of Biological Activity to Predict the Position and Stereochemistry of One of the Hydroxyl Groups. *Tetrahedron*, 43(13), 2985.

Karkhanis, D.W., Trivedi, G.K. and Bhattacharyya, S.C. (1978) Minor Sesquiterpene Alcohols of North Indian Vetiver Oil: Isolation and Structure of Isovalencenol, Vetiselinenol and Isovetiselinenol. *Ind. J. Chem.*, 16B, 260.

Kartha, C.C., Kalsi, P.S., Shaligram, A.M., Chakravarti, K.K. and Bhattacharyya, S.C. (1963) Terpenoids-XXXVIII. Structure and Stereochemistry of $(-)$-γ_2-Cadinene. *Tetrahedron*, 19, 241.

Kelly, R.B. and Eber, J. (1970) Total Synthesis of $(\pm)$-γ_2-Cadinene. *Can. J. Chem.*, 48, 2246.

Kelly, R.B. and Eber, J. (1972) A Total Synthesis of Cadin-4,10(15)-dien-3α-ol: Structure of Khusinol. *Can. J. Chem.*, 50, 3272.

Kido, F., Uda, H. and Yoshikoshi, A. (1967) The Structure of Zizanoic Acid, a Novel Sesquiterpene in Vetiver Oil. *Tetrahedron Letters*, 29, 2815.

Kido, F., Sakuma, R., Uda, H. and Yoshikoshi, A. (1969a) Minor Acidic Constituents of Vetiver Oil Part II. Cyclocopacamphenic and Epicyclocopacamphenic Acids. *Tetrahedron Letters*, 37, 3169.

Kido, F., Uda, H. and Yoshikoshi, A. (1969b) Total Synthesis of Zizaane-type Sesquiterpenoids. *J. C. S., Chem. Commun.*, 1335.

Kido, F., Uda, H. and Yoshikoshi, A. (1972) Synthetic Study on Zizaane-type Sesquiterpoids. *J. C. S., Perkin I*, 1755.

Kirtany, J.K. and Paknikar, S.K. (1971) North Indian Vetiver Oils: Comments on Chemical composition and Botanical origin. *Science and Culture*, 37(8), 395.

Kohli, J.C. (1976a) Complete Elimination of Khusol from Vetiver Oil. *Riechst., Aromen Koerperpflegem*, 26(9), 185.

Kohli, J.C. (1976b) Complete Elimination of Khusinol from Vetiver Oil. *Ann. Chim.*, 1(5), 247.

Kohli, J.C. (1977c) Specific Separation of Epikhusinol from Vetiver Oil by Thin-layer Chromatography. *Riechst., Aromen Kosmet.*, 27(3), 76.

Kohli, J.C. and Badaisha, K.K. (1985) Novel Synthesis of Epikhusinol. *Chem. Ind.*, 412.

Komae, H. and Nigam, I.C. (1968) Essential Oils and Their Constituents. XXXIX. Structures of Khusenic Acid and Isokhusenic Acid – Two Sesquiterpenic Constituents of Oil of Vetiver. *J. Org. Chem.*, 33, 1771.

Liu, H.-J. and Chan, W.H. (1979) A Total Synthesis of (-)-Khusimone. *Can. J. Chem.*, 57(6), 708.

Manchanda, S.K., Bhatia, I.S. and Kalsi, P.S. (1968) Chemical Study of Indian Vetiver Oil. I. *Riechst Aromen Koerperpflegem*, 18(10), 415, 418, 420, 422.

Manchanda, S.K., Bhatia, I.S. and Kalsi, P.S. (1970) Chemical Study of Indian Vetiver Oil. II. *Riechst Aromen Koerperpflegem*, 20(1), 3,4,6,8,10,12.

Marshall, J.A. and Andersen, N.H. (1967) The Structure of α-Vetivone (Isonootkatone). *Tetrahedron Letters.*, No. 17, 1611.

Marshall, J.A. and Pike, M.T. (1968) A Stereoselective Synthesis of α- and β-Agarofuran. *J. Org. Chem.*, 33, 435.

Marshall, J.A. and Brady, S.F. (1969) Stereochemical Relationships in Spirovetivane Sesquiterpenes: The Total Synthesis of Hinesol. *Tetrahedron Letters*, No. 18, 1387.

Marshall, J.A. and Brady, S.F. (1970a) The Total Synthesis of Hinesol. *J. Org. Chem.*, 35(12), 4068.

Marshall, J.A. and Johnson, P.C. (1970b) The Structure and Synthesis of β-Vetivone. *J. Org. Chem.*, 35, 192.

Marshall, J.A. and Warne, T.M. (Jr.) (1971) The Total Synthesis of (±)-Isonootkatone. Stereochemical Studies of the Robinson Annelation Reaction with 3-Penten-2-one. *J. Org. Chem.*, 36(1), 178.

Masada, Y. and Sumido, T. (1979) Composition of *Vetiveria* zizanioides Oil. 1. *Koen Yoshishu-Koryo, Terupen. oyobi Seiyu Kagaku ni Kansuru., Toronkai.*, 23, 209.

Maurer, von B., Fracheboud, M., Grieder, A. and Ohloff, G. (1972) 235. Zur Kenntnis der Sesquiterpenoiden C_{12}-Ketone des atherischen Ols Von *Vetiveria zizanioides* (L.) Nash. *Helv. Chim. Acta*, 55, 2371.

MacSweeney, D.F. and Ramage, R. (1971) A Stereospecific Total Synthesis of Zizanoic and Isozizanoic Acids. *Tetrahedron*, 27, 1481.

Mehta, G., Bhattacharyya, A. and Kapoor, S.K. (1973) Recent Progress in Sesquiterpene Synthesis: Part II – Tri- and Tetracyclic Sesquiterpenes. *J. Sci. Ind. Res.*, 32(4), 191.

Mizrahi, I. and Nigam, I.C. (1969) Essential Oils and Their Constituents. Isolation of Aromatic Sesquiterpenes from Reunion Vetiver Oil. *J. Pharm. Sci.*, 58(6), 738.

Naegeli, P. and Kaiser, R. (1972) A New Synthetic Approach to the Acorane-, Daucane- and Cedrane Skeleton. *Tetrahedron Letters*, No. 20, 2013.

Nair, E.V.G., Chinnamma, N.P. and Kumari, R.P. (1979) Review of the Work Done on Vetiver (*Vetivaria zizanioides Linn.*) at the Lemongrass Research Station, Odakkali. *Indian Perfumer*, 13(3 & 4), 199.

Nanda, J.K., Wadia, M.S. and Kalsi, P.S. (1970) Chemical Investigation of Indian Vetiver Oil. 3. *Riechst., Aromen, Koerperpflegem*, 20(5), 169.

Narain, K., Goswami, S. and Dasgupta, P.N. (1949) Chemical Examination of Indian Vetiver Oil. *Indian Soap J.*, 14(11), 303.

Neville, G.A. and Nigam, I.C. (1969) Long-range Coupling in the Novel Khusenate and Isokhusenate Series. *Tetrahedron Letters*, 10, 837.

Nigam, I.C., Dhingra, D.R. and Gupta, G.N. (1959) Potentiometric Estimation of Carbonyls in Indian Vetiver Oils. *Perfum. Essent. Oil Rec.*, 9(4), 297.

Nigam, I.C. and Levi, L. (1962) Gas Liquid Partition Chromatography of Sesquiterpene Compounds. *Can. J. Chem.*, **40**, 2083.

Nigam, I.C., Komae, H., Neville, G.A., Radecka, C. and Paknikar, S.K. (1968) Structural Relationships between Tricyclic Sesquiterpenes in Oils of Vetiver. *Tetrahedron Letters*, No. 20, 2497.

Nigam, I.C. and Komae, H. (1967) Essential Oils and Their Constituents XXXIV. Isolation of Khusenic and Isokhusenic acid from Oil of Vetiver and some Observations concerning Their Structures. *J. Pharm. Sci.*, **56**, 1299.

Nigam, I.C., Radecka, C. and Komae, H. (1968) Essential Oils and Their Constituents XXXVII. Isolation and Structure of Khusenol, a New Sesquiterpenic Primary Alcohol from Oil of Vetiver. *J. Pharm. Sci*, **57**, 1029.

Nguyen-Trong-Ann and Fetizon, M. (1965) The Sesquiterpene of Vetiver Oil. *Am. Perfumer Cosmet.*, **80**(3), 40,42,44.

Novotel'nova, N.F. and Rafanova, R. Ya. (1958) Composition of the Essential Oils of Vetiver. II. *Dushistykh Veshchestv*, No. 4, 85.

Novotel'nova, N.F., Rafanova, R. Ya. and Fuks, N.A. (1958) The Composition of Essential Oils of Vetiver. IV. *Trudy Vsesoyuz. Nauch.-Issledovatel. Inst. Sintet. i Natural. Dushistykh Veshchestv*, No. 4, 201.

Paknikar, S.K., Bhatwadekar, S.V. and Chakravarti, K.K. (1975) Biogenetically Significant Components of Vetiver Oil: Occurrence of (-)-α-Funebrene and Related Compounds. *Tetrahedron Letters*, No. 34, 2973.

Pfau, A.St. and Plattner, Pl. A. (1939) Volatile Plant Constituents. X. The Vetivones, odorous constituents of the Essential Oils of Vetiver. *Helv. Chim. Acta.*, **22**, 640.

Pfau, A.St. and Plattner, Pl. A. (1940) Volatile Plant Material. XI. The Constitution of β-Vetivone. *Helv. Chim. Acta.*, **23**, 768.

Raj, B., Kohli, J.C., Wadia, M.S. and Kalsi, P.S. (1971) Structure and Absolute Configuration of (+)-Khusitene, a Novel C_{14}-Hydrocarbon from Vetiver Oil. *Ind. J. Chem.*, **9**, 1047.

Rao, A.A., Surve, K.L., Chakravarti, K.K. and Bhattacharyya, S.C. (1963) Terpenoids-XXXVI. The Structure of Khusinol, a New Sesquiterpene Alcohol from Vetiver Oil. *Tetrahedron*, **19**, 233.

Sakuma, R. and Yoshikoshi, A. (1968) Tricyclovetivene. *J.C.S., Chem. Commun.*, 41.

Seshadri, R., Kalsi, P.S., Chakravarti, K.K. and Bhattacharyya, S.C. (1967) Terpenoids-XCIX. Structure and Absolute Configuration of Khusinoloxide, a New Antipodal Sesquiterpene Epoxy Alcohol from Vetiver Oil. *Tetrahedron*, **23**, 1267.

Shaligram, A.M., Rao, A.S. and Bhattacharyya, S.C. (1962) Terpenoids-XXXII. Absolute Configuration of Junenol and Laevojunenol and Synthesis of Junenol from Costunolide. *Tetrahedron*, **18**, 969.

Surve, K.L., Chakravarti, K.K. and Bhattacharyya, S.C. (1959) Chemical Examination of South Indian Vetiver Oil. *Chem. Ind.*, 1352.

Surve, K.L., Chakravarti, K.K. and Bhattacharyya, S.C. (1961) Chemical Examination of South Indian Vetiver Oil. *Indian Perfumer*, **5**(2), 135.

Takahashi, S. (1968) *Chem. Pharm. Bull.* (Japan), **16**, 2449.

Tirodkar, S.V., Paknikar, S.K. and Chakravarti, K.K. (1969) Khusinol: Location of Hydroxyl Group. *Science and Culture*, **35**(1), 27.

Tomita, B. (1971) *15th Symposium on Terpenes, Essential Oils and Aromatic Chemicals*, Osaka, Japan.

Trivedi, G.K., Chakravarti, K.K. and Bhattacharyya, S.C. (1971) *Ind. J. Chem.*, **9**, 1049.

Trivedi, G.K., Kalsi, P.S. and Chakravarti, K.K. (1964) Terpenoids-LIV. Structure and Absolute Configuration of Khusitone. *Tetrahedron*, **20**, 2631.

Trivedi, G.K. (1966) *Ph.D. Thesis*, Poona University.

Umarani, D.C., Gore, K.G. and Chakravarti, K.K. (1966) Terpenoids-XC. Khusimol, a New Sesquiterpene Alcohol. *Tetrahedron Letters*, No. 12, 1255.

Umarani, D.C., Gore, K.G. and Chakravarti, K.K. (1969) Terpenoids. CXXXVI. Khusimol –
a New Sesquiterpene Primary Alcohol from Vetiver Oil. *Perfumery Essent. Oil Rec.*, **60**, 307.

Umarani, D.C., Seshadri, R., Gore, K.G. and Chakravarti, K.K. (1970) *Flavour Ind.*, 1(9), 623.

Vig, O.P., Raj, I., Salota, J.P. and Matta, K.L. (1969) Terpenoids: Part XXXIII. Synthesis of
Isobisabolene. *J. Ind. Chem. Soc.*, 46(3), 205.

Vig, O.P., Sharma, S.D., Matta, K.L. and Sehgal, J.M. (1971) Terpenoids-Part LVIII. New
Synthesis of Iso-Bisabolene, β-Bisabolene and β-Terpineol. *J. Ind. Chem.Soc.*, 48(11), 993.

Yamada, K., Nagase, H., Hayakawa, Y., Aoki, K. and Hirata, Y. (1973) Synthetic Studies
on Spirovetivanes. II. Stereospecific Total Synthesis of *dl*-Hinesol, *dl*-α-vetispirene, and *dl*-
β-Vetispirene (*dl*-β-Isovetivenene). *Tetrahedron Letters*, No. 49, 4967.

Zutsi, N.L. and Sadgopal (1956) Fractionation of Indian Vetiver (Khus) Oil. *Perfum. Essent. Oil
Rec.*, 47(3), 88.

Zutsi, N.L. and Sadgopal (1957a) Sesquiterpenes from Indian Vetiver (Khus) Oil. *Perfum.
Essent. Oil Rec.*, 48(7), 333.

Zutsi, N.L. and Sadgopal (1957b) Physicochemical Examination of the Essential Oil from the
Roots of Vetiveria zizanioides. *Indian Soap J.*, 23(3), 43.

5 Ethnopharmacology and Pharmacological Properties of *Vetiveria Zizanioides*

Including Pharmacologic and Pharmacokinetic Properties

Nwainmbi Simon Chia

Belo Rural Development Project (BERUDEP), PMB, P.O. Box 5, Belo-Boyo, N.W. Province. Cameroon

Foreword to the chapter

Few information is available on the pharmacological properties of vetiver and vetiver oil. I though it was important to include, though limited, the personal Cameroonian experience of Dr Nwainmbi Simon Chia, who collected a certain amount of data that is presented here. Hence, this chapter has to be considered more as a short communication than a review chapter, being the only available data on vetiver ethnopharmacognosy and pharmacology.

Introduction

Pharmacology is the study of the action of drugs, substances influencing or attenuing the functioning of the body tissues, systems or organs on the living organism and potentially on human body. Ethnopharmacology refers specifically to the study of the action and use of primitive medicines derived usually from natural sources in tribal indigenous medicine, medicines that have stood the test of time and are still valued in tribal communities.

The medicinal value of the genus *Vetiveria* dates back to the 18th century with the migration of the Fulanis. They believed that when dry vetiver roots and leaves were burnt evil spirits and malpractices were sent away from where they settled. Later, they tried with some success to treat dental problems with the dried roots of the vetiver plant; however they soon forgot about the vetiver because the search for fertile pastures gradually caused nomadic movement away from areas where this grass was not found.

To suit the Cameroonian context of the vetiver grass, it was thought wise to call it "magic grass". In Thailand they call it the "miracle grass". In many countries they have their own common names for vetiver. The simple reasons for all these adjectives used for describing vetiver grass derive from its uses in ranging soil and water conservation, erosion control, soil stabilisation, for animal foods and for medicinal purposes.

Almost everyone suffers at some time from one or more health problems. Researchers on medicine have been working relentlessly to provide drugs that might solve all human health problems and even like the philosopher's stone maintain life forever.

To some it might seem that the number of new illnesses and health problems that are arising daily serve only to counteract the drugs produced. It is clear that a wide range of illnesses can be satisfactorily treated but the major problem is how can many people of the developing countries afford such drugs, and how long can they keep on thriving on very expensive drugs.

In BERUDEP, working hand in hand with other doctors of modern medicine we have found that the roots of *V. zizanioides* are capable of curing some common and dangerous health problems.

Today, though still not well established in well defined dose regimens, roots of *V. zizanioides* have been shown to take care of a variety of unrelated health hazards. Amongst such uses are antibiotherapy, antimalarial treatment, anti-inflammatory effects, and the treatment of stomatological and dietetic problems.

Processing

V. zizanioides roots should be harvested only between six and twelve calendar months after planting. These freshly harvested roots are washed and dried in a humid atmosphere at a controlled temperature between 22 °C and 36 °C for about two or three days. The roots are then stored in amber coloured polythene sachets. They should not be refrigerated at temperatures below 4 °C. Exposure to sunlight should be minimised to conserve the medicinal value of the roots. The roots are finally grounded into fine pellets and stored as above mentioned.

Differential Treatments

Antibiotic actions

For treatment, a dosage of the prepared powder was dispensed as follows; into 50 ml of fresh water a teaspoonful (= 3.9 g) of the prepared powder was introduced and boiled. After cooling, the filtrate obtained was given to the patient by the oral route. This was repeated three times a day for a period of seven days.

Reported cases of chronic Pelvic Inflammatory Diseases (PID) due to *Nesseria* and *Chlamydia* and to acute urinary tract infections caused by *Staphylococcus* (penicillinase and β-lactamase producers) and many other bacteria are treated with the roots of *V. zizanioides*.

Studies are being undertaken to establish the Maximum Inhibitory Concentration (MIC) of *V. zizanioides* as its roots can be a major help in some cases of septicaemia.

Antimalarial action

The commonest strain of malaria which was predominate in our area of work was *Plasmodium falciparum*. For treatment of malaria the same dosage and amount (i.e. 3.9 g of processed roots of *V. zizanioides* or processed leaves boiled and cooled in 50 ml fresh water) were used orally for a period of three days. There was some amelioration in the state of the patient. Much work needs to be done to check on the effectiveness of *V. zizanioides* on other haemoparasites such as Trypanosomes, Microfilariae and other species of *Plasmodium*.

Anti-inflammatory action

The active ingredients in the roots of *V. zizanioides* makes believe that it is a non-steriodal anti-inflammatory component. Two grams of the ground powder of the roots of *V. zizanioides* when chewed will relieve toothache in less than fifteen minutes. This treatment can be repeated as often as four times a day depending on the degree of pain.

Control of diabetes

Roots of *V. zizanioides* have a hypoglycaemic action. On consumption in two divided doses daily of one teaspoonful (3.9 g) in boiled 50 ml fresh water, the release of insulin from the pancreas was triggered and consequently reduced the Blood Sugar Level of a known insulin dependent diabetic. The drop in the Blood Sugar Level was very remarkable (14.1 mg/dl).

The patient has, as of today, been treated only with roots of *V. zizanioides* for the control of the Blood Sugar Level. This has greatly alleviated the inconvenience insulin administration causes as well as reducing the cost of control since *V. zizanioides* can be cultivated by the patient.

Lotion

V. zizanioides roots in a lotion prepared with bees wax is very effective in curing a wide range of skin diseases such as eczemas, insect bites, boils and other dermatological problems caused by deep seated fungal or bacterial infections.

Side Effects

To date no side effect or adverse reaction has been registered apart from cases of nausea due to overdose.

Shelf Life

When processed the roots of *V. zizanioides* are stable medicinally for a period of twelve calendar months at room temperature provided that they are stored away from direct sunlight and extremes of temperature.

Precaution and Contra-indications

The roots of *V. zizanioides* have not up to the present time shown any signs of causing renal insufficiency or blood dyscrasia. We do advise that roots of *V. zizanioides* should for now not be given to pregnant women, neonates and younger adults aged less than 12 years until well established rules are drawn up on completion of our study. We recommend patients on treatment with roots of *V. zizanioides* have their Blood Sugar Levels well controlled. Treatment should also be accompanied by adequate food.

Conclusion

The simple palliatives using the roots of *V. zizanioides* indicated in this chapter must not be used as a substitute for modern medical attention. For any serious health disturbances or for any chronic symptoms, we recommend the patient to seek proper medical attention from his or her physician or pharmacist.

6 Vetiver Grass Technology

Dr. Paul Truong

Principal Soil Conservationist, Resource Sciences Centre,
Queensland Department of Natural Resources. Brisbane, Australia.

Introduction

Although vetiver grass (*Vetiveria zizanioides L.*) has been used for land protection purposes in tropical and subtropical countries for more than a century, its real impact as a low cost, effective and simple method of soil conservation in agricultural land only emerged in the late 1980s. This was a result of its development for soil and water conservation by the World Bank in India and subsequently its promotion by The Vetiver Network, which was founded by Dick Grimshaw (see Chapter 1).

While its role in protecting agricultural lands remains very important, scientific research conducted in the last 10 years has clearly demonstrated that due to its special morphological and physiological attributes, vetiver grass provides unique opportunities for other applications such as infrastructure and environmental protection (Truong, 2000).

It is very important to point out that to obtain maximum effectiveness vetiver grass has to be used correctly for each application. For instance, as a method of soil and water conservation in agricultural land, in addition to engineering principles, a sound knowledge of soil and plant sciences are needed. Therefore the application of vetiver grass is called Vetiver Grass Technology (VGT).

Special Characteristics of Vetiver Grass

Morphological characteristics

Vetiver grass has no stolons and very short rhizomes and a massive finely structured root system that can grow very fast and in some applications rooting depth can reach 3–4 m in the first year. This deep root system makes the vetiver plant extremely drought tolerant and very difficult to dislodge in strong currents (Hengchaovanich, 1998).

Stiff and erect stems, which can stand up to relatively deep water flow (Truong *et al.*, 1995).

It is highly resistant to pests, diseases and fire (West *et al.*, 1996; Chen, 1999).

A dense hedge is formed when vetivers are planted close together acting as a very effective sediment filter and water spreader.

New shoots emerge from the base helping it to withstanding heavy traffic and heavy grazing pressure.

New roots are developed from nodes when the plant is buried by trapped sediment. Vetiver will continue to grow up despite the deposited silt, eventually forming terraces if the trapped sediment is not removed (Truong, 1999a).

Physiological characteristics

Valuable physiological characteristics include

- Tolerance to extreme climatic variation such as prolonged drought, flood, submergence and extreme temperature from −22 °C to 60 °C (Truong, 1999b; Xia *et al.*, 1999; Xu and Zhang, 1999);
- Ability to regrow very quickly after the weather improves or soil ameliorants have been added after having been affected by drought, frost, salinity and other adverse soil conditions (Truong *et al.*, 1995);
- Tolerance to a wide range of soil pH values (3.0 to 11.5) (Truong and Baker, 1998);
- High level of tolerance to herbicides and pesticides (Pithong *et al.*, 1996; Cull *et al.*, 2000);
- Highly efficient in absorbing dissolved nitrogen, phosphorus, mercury, cadmium and lead in polluted water (Pithong *et al.*, 1996; Suchada, 1996);
- Highly tolerant to growing media high in acidity, alkalinity, salinity, sodicity and magnesium (Truong and Baker, 1996; Truong and Baker, 1998) and highly tolerant to aluminium, manganese and heavy metals such as arsenic, cadmium, chromium, nickel, lead, mercury, selenium and zinc in the soils (Truong, 1999c).

Ecological characteristics

Although vetiver is very tolerant to some extreme soil and climatic conditions mentioned above, as a C4 plant (see Chapter 2) it is intolerant to shade. Shade will reduce its growth and in extreme cases may even eliminate vetiver in the long term. Therefore vetiver grows best in the open and weed control may be needed during the establishment phase. On erodible or unstable ground vetiver first reduces erosion, stabilising the erodible ground (particularly steep slopes), before improving its micro environment so that other wild or sown plants can establish later. Because of these characteristics vetiver can be considered as a pioneer plant on disturbed lands.

Weed potential

It is imperative that any plants used for bioengineering or environmental protection do not become weeds. To comply with the very strict Australian rules on introduced plants, a sterile vetiver cultivar was selected (from a number of existing cultivars in Australia) and exhaustively and rigorously tested for eight years for its sterility under various growing conditions. This cultivar is registered in Australia as Monto vetiver.

In Fiji, where vetiver grass was introduced to the country more than 100 years ago, and it has been widely used for soil and water conservation purposes for more than 50 years, yet vetiver grass has not become a weed in this new environment (Truong and Creighton, 1994).

As with all sterile plants, vetiver can only be vegetatively propagated and planted. Planting materials are obtained by subdividing the crown of a mature plant and are supplied as slips or splits in various forms suitable for different applications.

Although vetiver grass is very resilient under the most adverse conditions it can be eliminated easily either by spraying with glyphosate herbicide or uprooting and drying out by hand or using farm machinery.

Genetic characteristics

There are two vetiver species being used for soil conservation purposes, *Vetiveria zizanioides* L. and *V. nigritana*. The latter is native to southern and western Africa and its application is mainly restricted to that sub continent. As *V. nigritana* is a seeded variety its application should be restricted to its homeland because it may become a weed when introduced to new environments.

There are two *V. zizanioides* genotypes being used for soil and water conservation and for land stabilisation purposes:-

- The wild and seeded northern Indian genotype;
- The sterile or very low fertility southern Indian genotype.

While the seeded genotype is only used in northern India, the southern, sterile genotype, the vetiver used for essential oil production, is the genotype that is being used around the world for soil and water conservation and land stabilisation purposes. Results of the Vetiver Identification Program, by DNA typing, conducted by Adams and Dafforn (1997) have shown that of the 60 samples submitted from 29 countries outside South Asia, 53 (88%) were a single clone of *V. zizanioides*. These 53 samples tested came from North and South America, Asia, Oceania and Africa and most interestingly among these 53 cultivars are Monto (Australia), Sunshine (USA), and Vallonia (South Africa).

The implication is that, once the genotype is identified, all the research, development and application can be shared around the world. For example, as all vetiver research conducted in Australia has been based on Monto vetiver, all the Australian results published can be applied with confidence anywhere in the world when this genotype is used. A summary of the Monto vetiver adaptability range is shown on Table 6.1.

Main Applications of Vetiver Grass Technology

The main applications of VGT are based principally on some of the unique characteristics of vetiver grass mentioned above, namely:-

- its thick growth, forming dense hedges when planted in rows;
- its massive, fine, thick and deep root system and
- its high level of tolerance to adverse climatic and edaphic conditions and heavy metal toxicities.

Vegetative barriers as water spreaders and sediment filter

Hydraulic properties of vetiver hedges

When planted in rows, vetiver plants will form a thick hedge and with their stiff stems these hedges can stand up to water flow of at least 0.6 m depth, forming a living barrier which impedes and spreads run-off water. Hydraulic characteristics of vetiver hedges under deep flow were determined by flume tests at the University of Southern Queensland, Australia for flood mitigation on the flood plain of Queensland (Dalton *et al.*, 1996a) (Figure 6.1).

Table 6.1 Adaptability Range of Monto Vetiver in Australia and Other Countries.

Adverse Soil Conditions	Australia	Other Countries
Acidity	pH 3.3	pH 4.2 (with high level soluble aluminium)
Aluminium level (Al Sat. %)	Between 68%–87%	80%–87%
Manganese level	>578 mgkg^{-1}	
Alkalinity (highly sodic)	pH 9.5	pH 11.5
Salinity (50% yield reduction)	17.5 mScm^{-1}	
Salinity (survived)	47.5 mScm^{-1}	
Sodicity	33% (exchange Na)	
Magnesicity	2,400 mgkg^{-1} (Mg)	
Heavy Metals		
Arsenic	100–250 mgkg^{-1}	
Cadmium	20 mgkg^{-1}	
Copper	35–50 mgkg^{-1}	
Chromium	200–600 mgkg^{-1}	
Nickel	50–100 mgkg^{-1}	
Mercury	>6 mgkg^{-1}	
Lead	>1,500 mgkg^{-1}	
Selenium	>74 mgkg^{-1}	
Zinc.	>750 mgkg^{-1}	
Location	15°S–37°S	41°N–38°S
Climate		
Annual Rainfall (mm)	450–4,000	250–5,000
Frost (ground temp.)	–11 °C	–22 °C
Heat wave	45 °C	60 °C
Drought (without effective rain)	15 months	
Fertiliser		
Vetiver can be established on very infertile soil due to its strong association with mycorrhiza	N and P (300 kg/ha DAP)	N and P, farm manure
Palatability	Dairy cows, cattle, horse, rabbits, sheep, kangaroo	Cows, cattle, goats, sheep, pigs, carp
Nutritional Value	N = 1.1 % P = 0.17% K = 2.2%	Crude protein 3.3% Crude fat 0.4% Crude fibre 7.1%

Field trials using hydraulic characteristics determined by the above tests showed that vetiver hedges were successful in reducing flood velocity and limiting soil movement, resulting in very little erosion in fallow strips and a young sorghum crop was completely protected from flood damage.

Soil erosion and sediment control on sloping farmlands

As the vetiver hedge is a living porous barrier it slows and spreads run-off water and traps sediment. As water flow is slowed down, its erosive power is reduced and at the

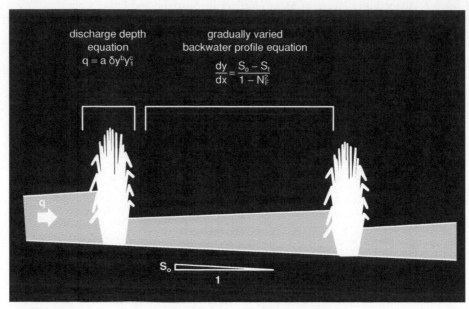

discharge depth
equation
$q = a\,\delta y^b y_i^c$

gradually varied
backwater profile equation

$$\frac{dy}{dx} = \frac{S_o - S_f}{1 - N_F^2}$$

S_o

1

q = discharge per unit width   y = depth of flow y_1 = depth upstream
S_o = land slope S_f = energy slope N_F = the Froude number of flow.

Figure 6.1 Hydraulic model of flooding through vetiver hedges.

same time allows more time for water to infiltrate to the soil, and the hedge traps any eroded material. Therefore an effective hedge will reduce soil erosion, conserve soil moisture and trap sediment on site. When appropriately laid out these hedges can also act as very effective diversion structures spreading and diverting run-off water to stable areas or proper drains for safe disposal.

This is in sharp contrast to the stratagem with the contour terrace/waterway system in which run-off water is collected by the terraces and diverted as quickly as possible from the field to reduce its erosive potential. All this run-off water is collected and concentrated in the waterways where most erosion occurs in agricultural lands, particularly on sloping lands and this water is lost from the field. With the VGT not only is this water conserved but also no land is wasted on troublesome artificial waterways.

Both research findings and field results in Australia, Asia, Africa and South America show that in comparison with conventional cultivation practices, surface run-off and soil loss from fields treated with vetiver were significantly lower and crop yield was much improved by as much as 20%. The yield increase was attributed mainly to uniform *in situ* soil and moisture conservation over the entire toposequence under the vetiver hedge system (Truong, 1993).

Erosion and sediment control on floodplain

VGT has been used as an alternative to strip cropping practice on the flood plain of Queensland, Australia. This practice relies on the stubble of previous crops for erosion control of fallow land and young crops. On an experimental site vetiver hedges that

Figure 6.2 This 2000 m long vetiver hedge was established to protect a sorghum crop from flood damage on the flood plain of the Darling Downs, Australia.

were established at 90 m intervals provided a permanent protection against flooded water. Results over the last several years (including several major flood events) have shown that VGT is very successful in reducing flood velocity and limiting soil movement with very little erosion in the fallow strips (Figure 6.2) (Dalton *et al.*, 1996a; Dalton *et al.*, 1996b; Truong *et al.*, 1996a).

The incorporation of vetiver hedges as an alternative to strip cropping on floodplains has resulted in more flexibility, more easily managed land and more effective spreading of flood flows in dry years particularly with low stubble producing crops such as cotton and sunflower. An added major benefit is that the area cropped at any one time could be increased by up to 30% (Dalton *et al.*, 1996a).

Rehabilitation Saline and Acid Sulfate Soils

The spread of salinity in both dryland and irrigated lands is a major concern in low rainfall and semiarid regions of the world. Vetiver has been used very successfully in erosion control and rehabilitation of these salt-affected lands (Truong, 1994).

Acid Sulfate Soils (ASS) constitute a major component of arable lands in many tropical countries in Africa and Asia such as Thailand and Vietnam where rice is the main food crop. These soils are highly erodible and difficult to stabilise and rehabilitate. Eroded sediment and leachate from ASS are extremely acidic. The leachate from ASS has led to disease and death of fish in several coastal zones of eastern Australia. Vetiver has been successfully used to stabilise and rehabilitate a highly erodible acid sulfate soil on the coastal plain in tropical Australia, where actual soil pH is around 3.5 and oxidised pH is as low as 2.8 (Truong and Baker, 1996, 1998).

Infrastructure protection

The major impact of VGT in the late 1990s was in the area of steep slope stabilisation for infrastructure protection (Hengchaovanich, 1999; Xie, 1997; Xia *et al.*, 1999).

Batter stabilisation

The main causes of slope instability are surface erosion and structural weakness of the slope. While surface erosion often leads to rill and gully erosion, structural weakness will cause mass movement or land slip.

Research conducted by Hengchaovanich and Nilaweera (1996) in Malaysia showed that the tensile strength of vetiver roots increases with the reduction in root diameter; this phenomenon implies that stronger fine roots provide higher resistance than larger roots. The tensile strength of vetiver roots varies between 40–180 Mpa for the range of root diameters between 0.2–2.2 mm. The mean design tensile strength is about 75 Mpa (equivalent to approximately one sixth of mild steel) at 0.7–0.8 mm diameter, which is the most common size for vetiver roots. This indicates that vetiver roots are as strong as, or even stronger than roots of many hardwood species, which have been proven positive for root reinforcement in steep slopes (Figure 6.3).

In a soil block shear test, Hengchaovanich and Nilaweera (1996) also found that root penetration of a two year old vetiver hedge with 15 cm plant spacing can increase the shear strength of soil in the adjacent 50 cm wide strip by 90% at 0.25 m depth. The increase was 39% at 0.50 m depth and gradually reduced to 12.5% at 1 m depth. Moreover, because of its dense and massive root system it offers better shear strength increase per unit fibre concentration (6–10 kPa/kg of root per cubic metre of soil compared to 3.2–3.7 kPa/kg for tree roots).

The major advantages of VGT over conventional engineering methods in infrastructure protection are:–

Figure 6.3 Vetiver Grass Technology is used as a bioengineering technique to stabilise this very steep cut batter of a road in the Philippines where landslides often occurred during the annual typhoon season. (Photo Credit: Noah Manarang).

- VGT provides a "soft" option using a green and natural method of stabilisation rather than the "hard" conventional engineering approach such as rocks and concrete structures;
- VGT is very cost effective when compared with the costs of conventional methods. VGT costs only 10–17% of the cost of rocks in country with low labour cost as in China (Xia et al., 1999) and between 27–40% in a country with high labour costs such as Australia (Bracken and Truong, 2000).

Landslide remediation

Landslides are often caused by the lack of structural strength of the ground on steep slopes and the event is triggered by over-saturation during heavy rainfall periods. The problem can be exacerbated by the presence of tall trees, which can be blown over by strong wind. Under natural conditions such as forests, deep rooted trees provide the structural reinforcement, but when deforestation was carried out for agriculture and forestry production or infrastructure construction this structural reinforcement was lost. When this occurred, landslide often resulted. Vetiver grass with its extensive root system provides the much-needed soil reinforcement in a much shorter time than trees. In addition, vetiver grass has no heavy canopy to blow over in strong wind (Miller, 1999).

Floods mitigation

The combination of the deep root system and thick growth of the vetiver hedges will protect the banks of rivers and streams under flood conditions. Its deep roots prevent it from being washed away while its thick top growth reduces flow velocity and its erosive power. In addition, when properly laid out hedges can be designed to direct water flow to appropriate areas (Truong, 1999b).

Very successful stream bank and riverbank stabilisation has been carried out in Australia, Malaysia and the Philippines.

Land rehabilitation

Mine rehabilitation

On site and offsite pollution control from mining wastes is a major application of VGT for environmental protection. Research conducted by this author has clearly established the extremely high levels of tolerance of vetiver grass aluminium, manganese and heavy metals such as arsenic, cadmium, chromium, nickel, copper, lead, mercury, selenium and zinc in the soils (Table 6.2) (Truong and Baker, 1998).

Table 6.3 shows that distribution of heavy metals in vetiver plant can be divided into three groups:-

- very little of the arsenic, cadmium, chromium and mercury absorbed was translocated;
- a moderate proportion of copper, lead, nickel and selenium was translocated (16% to 33%) and
- zinc was almost evenly distributed between shoot and root (40%).

Table 6.2 Threshold Levels of Heavy Metals to Vetiver Growth.

Heavy Metals	Thresholds to Plant Growth ($mgKg^{-1}$)		Thresholds to Vetiver Growth ($mgKg^{-1}$)	
	Hydroponic levels	Soil levels	Soil levels	Shoot levels
Arsenic	0.02–7.5	2.0	100–250	21–72
Cadmium	0.2–9.0	1.5	20–60	45–48
Copper	0.5–8.0	NA	50–100	13–15
Chromium	0.5–10.0	NA	200–600	5–18
Lead	NA	NA	>1,500	>78
Mercury	NA	NA	>6	>0.12
Nickel	0.5–2.0	7–10	100	347
Selenium	NA	2–14	>74	>11
Zinc	NA	NA	>750	880

NA not available.

Table 6.3 Average Distribution of Heavy Metals in Vetiver Shoots and Roots.

Metals	Soil ($mgKg^{-1}$)	Shoot ($mgKg^{-1}$)	Root ($mgKg^{-1}$)	Shoot/Root %	Shoot/Total %
Arsenic	688.4	8.4	180.2	4.8	4.6
Cadmium	1.0	0.3	11.0	3.1	2.9
Copper	50	13	68	19	16
Chromium	283.3	9.0	1,108	<1	<1
Lead	469	35	46	57	33
Mercury	1.98	0.05	2.27	6	5
Nickel	300	448	1,040	43	30
Selenium	19.9	4.4	8.4	53	33
Zinc	390	461	643	69	40

The important implications of these findings are that when vetiver is used for the rehabilitation of sites contaminated with high levels of arsenic, cadmium, chromium and mercury, it can be safely grazed by animals or harvested for mulch, as very little of these heavy metals are translocated to the shoots. As for copper, lead, nickel, selenium and zinc their uses for the above purposes are limited to the thresholds set by the environmental agencies and the tolerance of the animal concerned.

In addition, although vetiver is not a hyper-accumulator it can be used to remove some heavy metals from contaminated sites and disposed off safely elsewhere, thus gradually reducing the contaminant levels. For example, vetiver roots and shoots can accumulate more than five times the chromium and zinc levels in the soil.

These results have led the two major mining countries, Australia and South Africa, to increasingly adopting VGT as a major component of their rehabilitation strategy (Truong, 1999c).

In Australia VGT is highly successful in the rehabilitation of old quarries and mines where very few species can be established due to the hostile environment. Vetiver is able to stabilise the erodible surface first so that other species can colonise the areas between the hedges later. After two years the site is completely revegetated with vetiver and local species (Truong *et al.*, 1995). In Queensland, vetiver has been successfully used to stabilise mining overburden and highly saline, sodaic and alkaline tailings of coal mines (Radloff *et al.*, 1995) and highly acidic (pH 3.5) tailings of a gold mine. Recently VGT has been used to rehabilitate bentonite mine wastes and tailings from dam walls of major bauxite and copper mines and alumina refineries in northern Australia (Figure 6.4) (Truong, 1999c; Bevan and Truong, 2000).

In South Africa, rehabilitation trials conducted by De Beers on both tailing dumps and slimes dams at several different sites have shown that vetiver possesses the necessary attributes for self-sustainable growth on kimberlite spoils (Knoll, 1997). Vetiver grew vigorously on kimberlite, retaining run-off, arresting erosion and creating an ideal micro-habitat for the establishment of indigenous grass species. At Premier (800 mm annual rainfall) and Koffiefonteine (300 mm rainfall) diamond mines the surface temperature of the black kimberlite often exceeds 55 °C and most seeds are unable to germinate at this temperature. Vetiver planted at 2 m VI provided shade that cooled the surface and allowed germination of other grass seeds (Grimshaw, *pers. com.*). More recently very successful rehabilitation of slimes dams has also been carried out at Foskor mines (Tantum, *pers. com.*).

Landfill and contaminated lands rehabilitation

Old landfill and industrial waste dumps such as tanneries, galvanised factories and electrolytic factories are usually contaminated with heavy metals such as arsenic, cadmium, chromium, mercury, lead and zinc. As these heavy metals are highly toxic to human beings, the movement of these metals off-site must be controlled.

Rehabilitation works were carried out at an old landfill site at Cleveland in Australia by planting vetiver rows on the side slopes for erosion control. For leachate control, vetiver was planted *en masse* at the toe of the slope where leachate appeared. Although the landfill was heavily contaminated, vetiver established easily and grew well with nitrogen and phosphorus application at planting. The slopes were completely stabilised within 12 months and local vegetation established naturally between the rows. During the same period, leachate export was reduced substantially during the wet season and was eliminated during the dry season (Truong *et al.*, 1996b). When the slope was stabilised, native trees and shrubs were planted to complete the rehabilitation work. In this application vetiver acted as a pioneer plant (Figure 6.5).

Waste water treatment

Purification of polluted water

In China research results showed that vetiver can reduce soluble P up to 99% after 3 weeks and 74% of soluble N after 5 weeks. With proper planning VGT has the potential for removing up to 102 ton of nitrogen and 54 ton of phosphorus per year per hectare of vetiver planting (Zheng *et al.*, 1997).

Control of algal growth in rivers and dams

As soluble nitrogen and particularly soluble phosphorus are usually considered to be key elements for water eutrophication which normally leads to blue green algal growth

Figure 6.4 Vetiver is used to stabilise the deeply cut gullies of the overburden stockpile of an old coal mine (background) in Queensland, Australia. The horizontal rows in the foreground were planted to trap eroded material from moving offsite.

Figure 6.5 A thick stand of vetiver was planted on the side slope of this old landfill to soak up leachate run-off, preventing it from contaminating the surrounding environment.

in inland waterways and lakes, the removal of these elements by vegetation is a most cost effective and environmental friendly method of controlling algal growth.

Chinese works indicated that vetiver could remove dissolved nutrients and reduced algal growth within two days under experimental conditions (Xia *et al.*, 1997). Therefore, VGT can be used very effectively to control algal growth in water infested with blue-green algae. This can be achieved by planting vetiver on the edges of the streams or in the shallow parts of the lakes where usually high concentrations of soluble nitrogen and phosphorus occurred. Alternatively vetiver can be grown hydroponically on floating platforms which can be moved to the worst affected parts of the lake or pond. The advantages of the platform method is that vetiver tops can be harvested easily for stock feed, mulch or paper production and vetiver roots can also be removed for essential oil production (Xu and Zhang, 1999).

Effluent disposal

With the potential of removing very high quantities of nitrogen and phosphorus, vetiver planting can be used to remove nitrogen, phosphorus and other nutrients in effluent from sewage, abattoirs, feedlots, piggeries and other intensive livestock industries. In Australia VGT was used very successfully as an integral part of a waste water purification program in removing nutrients from effluent from septic tanks.

Trapping agrochemicals and nutrients

When established across drainage lines and watercourses, vetiver hedges filter and trap both coarse and fine sediment resulting in cleaner run-off water.

In Australia the combination of vetiver and African star grass has filtered out both the bed and the suspended load of run-off water on a pineapple farm in Queensland. Sediment load was reduced from 3.94 g/L to 2.33 g/L after passing through the hedge, and similarly, electrical conductivity was reduced to half (263 uSm/cm to 128 uSm/cm). The high dose of weedicide used did not affect vetiver growth (Ciesiolka, 1996). In other trials on sugarcane and cotton farms, vetiver hedges planted across drainage lines were particularly efficient in trapping particulate sediment containing high concentration of nutrients and agrochemicals. In sugar cane farms 69% of P in run-off sediments were trapped (Table 6.4) and on a cotton farm from 67% to 90% of pesticides, 48% of herbicides, 52% of P, 73% of N and 55% of S were trapped (Figures 6.6 and 6.7) (Truong et al., 2000).

Wetland application

In addition to the high levels of tolerance of vetiver towards agrochemicals and heavy metals mentioned above, vetiver thrives under wet or water-logged conditions. Therefore it is highly suitable for use in the wetland system to remove pollutants from industrial as well as agricultural wastes such as nutrients and agrochemicals from polluted water discharged from cropping lands and aquaculture ponds.

Recent research in Australia also demonstrated that vetiver can tolerate extremely high levels of Atrazine and Diuron under wetland conditions or in water bodies downstream. Indeed, vetiver was shown to be unaffected by either herbicide at concentrations as high as 2,000 μg L^{-1}, levels which are likely to be encountered in the environment only in situations of accidental spillage or direct application to waterways (Figure 6.8). Vetiver's tolerance towards the two herbicides was demonstrated in terms of both growth (water use, cumulative leaf area and dry weight) and photosynthetic activity (PAM fluorometry). At the concentration of 2,000 μg/L, other wetland plants, including *Phragmites australis* were either killed or their growth was severely reduced (Figure 6.9). These chemicals are the two most commonly used weedicides in the sugarcane industry in Australia (Cull et al., 2000).

Planting Materials and their Suitability

As mentioned earlier, vetiver grass has to be established vegetatively by subdivision of the crown, yielding slips. Each slip normally consists of 2–3 actively growing tillers. In Australia, four types of planting material are being used:–

- *Bare-root slips* are freshly subdivided splits from large and mature clumps of vetiver grass. Although these slips are the cheapest to use, they are for immediate planting only and they require most intensive watering during hot and dry periods. The main disadvantages of the bare-root slips are that they have high mortality rate and slower growth following planting. Therefore it is recommended only for small projects or agricultural applications where watering can be done easily.

- *Bare-root plantlets* are bare-root slips that have been raised in sand beds or containers for 5–6 weeks. These are small young plants with 3 to 4 tillers and a well developed root system; they are supplied bare root and suitable for planting within a week. In large projects, these plantlets can be raised on site to reduce costs. Although the plantlets have a better establishment rate than bare root slips,

Table 6.4 Nutrient Concentrations in Sediment Collected From Various Treatments.

| | Treatments | | | Analytical Results | | | | | Mg | | | Org. |
Soil Surface	Trash Cover	Fertiliser Placement	Vetiver Hedges	pH	Total N %	Bicarb P mg/kg	K	Ca	cmol(+)/kg	Na	ECEC	C %
Rotary hoe	Nil	Buried	NO	7.05	0.09	34.5	0.10	1.36	0.85	0.04	2.34	1.10
Rotary hoe	Nil	Buried	Yes	6.65	0.07	11.5	0.05	0.66	0.42	0.02	1.15	0.80
Zero till	Burnt trash blanket	Buried	NO	6.55	0.08	18.0	0.08	0.95	0.54	0.07	1.64	0.80
Zero till	Burnt trash blanket	Buried	Yes	6.95	0.08	13.0	0.09	0.74	0.46	0.03	1.31	0.75
Zero till	Green Trash blanket	Surface	NO	7.00	0.95	35.5	0.10	0.72	0.50	0.07	1.39	0.85
Zero till	Green Trash blanket	Surface	Yes	7.10	0.03	11.0	0.01	0.31	0.36	0.03	0.71	0.30
Original soil (0–0.25 m)				5.5	1.50	13.0	1.5	0.50	0.60	0.07	0.71	0.30

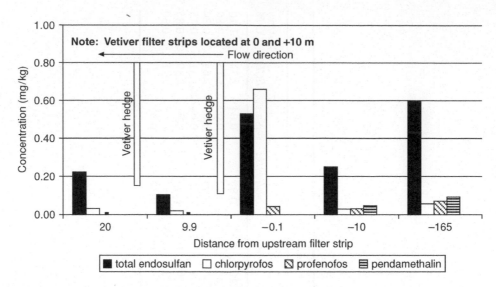

Figure 6.6 Pesticide concentration in deposited soil upstream and downstream of vetiver filter strips planted on the drainage line of a cotton farm in Australia.

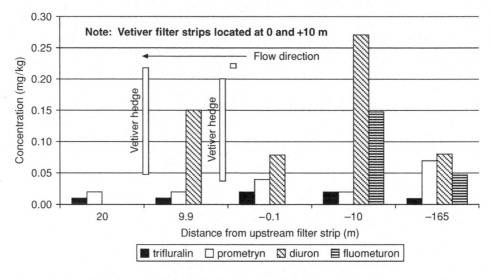

Figure 6.7 Herbicide concentration in deposited soil upstream and downstream of vetiver filter strips planted on the drainage line of a cotton farm in Australia.

like the bare-root slips, their initial growth is much slower than other types of planting material mentioned below. The plantlets are suitable for airfreight and for machine planting but they also need intensive watering during the establishment phase. They are recommended for large scale on site application where good watering is available.

- *Tubed stocks* are tubed or potted plants (6–8 week old) which are young plants with at least 3 medium size tillers and a well-developed root system. Although

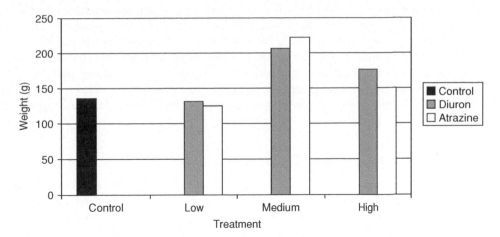

Figure 6.8 Mean effect of 3 levels of atrazine and diuron on the whole plant dry weight of vetiver at harvest, NS ($P \pounds 0.05$).

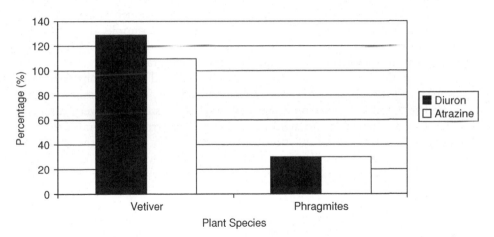

Figure 6.9 Comparison of mean whole plant dry weights of vetiver and *Phragmites* at high rates of herbicide application, when expressed as a percentage of dry weights of respective controls.

tubed plants are more expensive, they have the advantage over the two kinds of planting materials mentioned above in that they can continue to grow uninterrupted following planting. This will ensure good establishment and the fast growth that is vital for bioengineering and rehabilitation projects. Another advantage with tubed stocks is that they can be kept on site for a longer period and planted out when needed. Tube stocks have the best establishment survival rate and require less watering during the establishment phase and are therefore recommended for large infrastructure stabilisation and mine rehabilitation projects or sites where watering is difficult or impractical such as on high slopes.

- *Strips* are bands of vetiver in various lengths, which were raised in special containers for 2–3 months. In addition to the advantages mentioned for tubed stock, the main advantage of the strips is that the vetiver plants are established together (50–70 mm apart) and root damage is minimal during planting. The other advantage is the lower planting costs as strips are faster and easier to plant (up to 1 m length a time) especially on steep slopes. Because of the smaller gaps between plants these strips provide protection sooner than other planting materials. The strips also require less intensive watering, but their main disadvantage is the slightly higher costs. It is highly recommended for sediment filtration and water flow control works.

Appropriate Designs and Techniques

It should be stressed that VGT is a new technology. As any new technology it has to be learnt and applied appropriately for best results. Failure to do so will bring disappointing outcomes and sometimes adverse results. As a soil conservation technique and recently a bioengineering tool, the application of VGT requires the understanding of biology, soil science and also hydraulic and hydrological principles.

In addition, it has to be understood that vetiver is a grass by botanical classification but it acts more like a tree than a typical grass with its stiff, erect shoots and extensive, deep root system. To add to the confusion, as mentioned above, VGT exploits its different characteristics for different applications; for example, deep roots for land stabilisation, thick growth for water spreading and sediment trapping and extraordinary tolerance to various chemicals for land rehabilitation etc.

Failures of VGT in most cases can be attributed to bad applications rather than the grass itself or the technology recommended. For example, in one instance when vetiver was used to stabilise batters on a new highway, the results were very disappointing and failures to establish or to stabilise the slopes occurred at several sites along the highway. It was later found out from the engineers who specified the VGT, that the nursery that supplied the planting materials to the field supervisors and labourers, who planted the vetiver, had no previous experience or training in the use of VGT for steep slopes stabilisation.

Conclusion

From the research results and the successes of numerous applications presented above, it is clear that VGT provides a very effective and low cost environmental protection tool for the stabilisation of steep slopes, rehabilitation of contaminated land and purification of polluted water.

However it must be emphasised that to achieve the satisfactory results, the three most important factors are a) *good quality planting material*, b) *the all-important appropriate design* and c) *correct planting techniques*.

References

Adams, R.P. and Dafforn, M.R. (1997) DNA fingerprints (RAPDs) of the pantropical grass, *Vetiviria zizanioides L*, reveal a single clone, "Sunshine," is widely utilised for erosion control. *The Vetiver Newsletter*, 18, 28–33. Leesburg Va, USA.

Bevan, O. and Truong, P. (2000) The effectiveness of vetiver grass in erosion and sediment control at a Bentonite mine in Queensland, Australia. *Proc. Second International Vetiver Conf.*, Thailand, January 2000 (in press).

Bracken, N. and Truong, P.N. (2000) Application of Vetiver Grass technology in the stabilisation of road infrastructure in the wet tropical region of Australia. *Proc. Second International Vetiver Conf.*, Thailand, January 2000 (in press).

Chen, Shangwen (1999) Insect on vetiver hedges. *The Vetiver Newsletter*, **20**, 30–31 Leesburg Va, USA.

Ciesiolka, C.A. (1996) Vetiver grass as a component in a steep land farming system in south east Queensland. *Proc. Workshop on Research, Development and Application of Vetiver Grass for Soil Erosion and Sediment Control in Queensland.* Toowoomba, Australia, November 1996.

Cull, H., Hunter, H., Hunter, M. and Truong, P. (2000) Application of VGT in off-site pollution control. II-Tolerance of vetiver grass towards high levels of herbicides under wetland conditions. *Proc. Second International Vetiver Conf.*, Thailand, January 2000 (in press).

Dalton, P.A., Smith, R.J. and Truong, P.N.V. (1996) Hydraulic characteristics of vetiver hedges: An engineering design approach to flood mitigation on a cropped floodplain. *Proc. First Intern. Vetiver Conf.*, Thailand, February 1996.

Dalton, P.A., Smith, R.J. and Truong, P.N.V. (1996) Vetiver grass hedges for erosion control on a cropped flood plain: Hedge Hydraulics. *Agric. Water Management*, **31**, 91–104.

Hengchaovanich, D. (1998) Vetiver grass for slope stabilisation and erosion control, with particular reference to engineering applications. *Technical Bulletin No. 1998/2. Pacific Rim Vetiver Network.* Office of the Royal Development Projects Board, Bangkok, Thailand.

Hengchaovanich, D. (1999) Fifteen years of bioengineering in the wet tropics from A *(Acacia auriculiformis)* to V *(Vetiveria zizanioides). Proc. Ground and Water Bioengineering for Erosion Control and Slope Stabilisation*, Manila, April 1999.

Hengchaovanich, D. and Nilaweera, N.S. (1996) An assessment of strength properties of vetiver grass roots in relation to slope stabilisation. *Proc. First Intern. Vetiver Conf.*, Thailand, February 1996.

Knoll, C. (1997) Rehabilitation with Vetiver. *African Mining Magazine*, **2**, 43.

Miller, J. (1999) Experience with the use of vetiver for the protection and stabilisation of infrastructure. *Report on Bioengineering Workshop for Post-Hurricane Mitch Construction, San Salvador*, Latin America Vetiver Network, El Salvador, May 1999.

Pithong, J., Impithuksa, S. and Ramlee, A. (1996) Capability of vetiver hedgerows on the decontamination of agro chemicals residues. *Proc. First Intern. Vetiver Conf.*, Thailand, February 1996.

Radloff, B., Walsh, K. and Melzer, A. (1995) Direct revegetation of coal tailings at BHP Saraji Mine. *Aust. Mining Council Envir. Workshop*, Darwin, Australia, 1995.

Suchada, K. (1996) Growth potential of vetiver grass in relation to nutrients in wastewater of Changwat Phetchubari. *Proc. First Intern. Vetiver Conf.*, Thailand, February 1996.

Truong, P.N.V. (1993) Report on the international vetiver grass field workshop, Kuala Lumpur. *Australian Journal of Soil and Water Conservation*, **6**, 23–26.

Truong, P.N.V. (1994) Vetiver grass, its potential in the stabilisation and rehabilitation of degraded and saline lands. In V.R. Squire and A.T. Ayoub (eds.), *Halophytes a resource for livestock and for rehabilitation of degraded land*, Kluwer Academics Publisher, Netherlands, pp. 293–296.

Truong, P.N.V. (1999a) Vetiver Grass Technology for land stabilisation, erosion and sediment control in the Asia Pacific region. *Proc. Ground and Water Bioengineering for Erosion Control and Slope Stabilisation*, Manila, April 1999.

Truong, P.N.V. (1999b) Vetiver Grass Technology for flood and stream bank erosion control. *Proc. Intern. Vetiver Workshop*, Nanchang, China, October 1999 (in press).

Truong, P.N.V. (1999c) Vetiver Grass Technology for mine tailings rehabilitation. *Proc. Ground and Water Bioengineering for Erosion Control and Slope Stabilisation*, Manila, April 1999.

Truong, P.N. (2000) The Global Impact of Vetiver Grass Technology on the Environment. *Proc. Second Intern. Vetiver Conf.* Thailand, January 2000 (in press).

Truong, P. and Creighton, C. (1994) Report on the potential weed problem of vetiver grass and its effectiveness in soil erosion control in Fiji. *Division of Land Management*, Queensland Department of Primary Industries, Brisbane, Australia.

Truong, P., Baker, D. and Christiansen, I. (1995) Stiffgrass barrier with vetiver grass – A new approach to erosion and sediment control. *Proc. Third Annual Conference on Soil and Water Management for Urban Development*, Sydney, Australia, 1995.

Truong, P.N. and Baker, D. (1996) Vetiver grass for the stabilisation and rehabilitation of acid sulfate soils. *Proc. Second National Conf. Acid Sulfate Soils*, Coffs Harbour, Australia, 1996.

Truong, P.N. and Baker, D. (1997) The role of vetiver grass in the rehabilitation of toxic and contaminated lands in Australia. *Proc. Intern. Vetiver Workshop*, Fuzhou, China, October 1997.

Truong, P.N. and Baker, D. (1998) Vetiver Grass System for environmental protection. *Technical Bulletin No. 1998/1. Pacific Rim Vetiver Network*, Office of the Royal Development Projects Board, Bangkok, Thailand.

Truong, P.N.V., Dalton, P.A., Knowles-Jackson, C. and Evans, D.S. (1996a) Vegetative Barrier with Vetiver grass: An alternative to conventional soil and water conservation systems. *Proc. 8th Australian Agronomy Conf.*, Toowoomba, Australia, February 1996.

Truong, P., Baker, D. and Stone, R. (1996b) Vetiver grass for the stabilisation and rehabilitation of contaminated lands. Poster paper, *Workshop on Research, Development and Application of Vetiver Grass for Soil Erosion and Sediment Control in Queensland*, Toowoomba, Australia, November 1996.

Truong, P., Mason, F., Waters, D. and Moody, P. (2000) Application of VGT in off-site pollution control. I-Trapping agrochemical and nutrients in agricultural lands. *Proc. Second Intern. Vetiver Conf.*, Thailand, January 2000 (in press).

West, L., Sterling, G. and Truong, P.N. (1996) Resistance of vetiver grass to infection by root-knot nematodes (*Meloidogyne spp*). *The Vetiver Network Newsletter*, **20**, 20–22, Leesburg, VA, USA.

Xia Hanping, Ao Huixiu, Lui Shizhong and He Daoquan (1997) A preliminary study on vetiver's purification for garbage leachate. *Proc. Intern. Vetiver Workshop*, Fuzhou, China, October 1997.

Xia, H.P., Ao, H.X., Liu, S.Z. and He, D.Q. (1999) Application of the vetiver grass bioengineering technology for the prevention of highway slippage in southern China. *Proc. Ground and Water Bioengineering for Erosion Control and Slope Stabilisation*, Manila, April 1999.

Xie, F.X. (1997) Vetiver for highway stabilisation in Jian Yang County: Demonstration and Extension. *Proc. Intern. Vetiver Workshop*, Fuzhou, China, October 1997.

Xu, L. and Zhang, J. (1999) An overview of the use of vegetation in bioengineering in China. *Proc. Ground and Water Bioengineering for Erosion Control and Slope Stabilisation*, Manila, April 1999.

Zheng ChunRong, Tu Cong and Chen Huai Man (1997) Preliminary experiment on purification of eutrophic water with vetiver. *Proc. Intern. Vetiver Workshop*, Fuzhou, China, October 1997.

7 Biotechnology

Marco Mucciarelli[1] *and Ruth E. Leupin*[2]

[1] Department of Veterinary Morphophysiology, University of Turin, Viale P.A. Mattioli 25, I–10125 Turin, Italy.
[2] Institute of Biotechnology ETH Hönggerberg, CH–8093 Zürich, Switzerland.

Introduction

There is an urgent need to preserve crop plants through improved methods of vegetative propagation and this is especially true with tropical crops. Successful agricultural practice requires that many of the tropical species be reproduced vegetatively. In addition many lines that are or may be valuable as gene pools for breeding are lost each year by neglect or by mass eradication of their habitats for alternative uses.

Selected germlines of *Vetiveria zizanioides* (L.) Nash have long been cultivated for their odorous roots, which contain the essential oil of vetiver, used extensively in perfumery and cosmetics. The principal oil producing regions are Java, Brazil, Reunion Islands, Haiti, Angola, China, Guatemala and India.

From a morphological point of view, it is not possible to distinguish among different vetiver races or varieties, except for the well known existence of two broad types, the North India Type, known as "Khus" varieties of vetiver oil, consisting of the wild, fertile population native to the Ganges plain from Pakistan to Bangladesh and the South India Type, the "hedgerow" vetiver, which is sterile and cultivated traditionally for its essential oil content (Peyron, 1995; Adams and Dafforn, 1997a). This oil is clearly distinguished physicochemically (dextrorotation) in commerce from Khus oil, and it has long been produced pantropically through vetiver cuttings.

The exact origin of "hedgerow" non-seeding vetiver is unknown; however, *Vetiveria zizanioides* seems to have originated in the area stretching from India to Vietnam. North India type populations show a reluctant tendency to self inbreed so that little interest has been devoted by growers to the screening and selection of new germlines. This fact, together with low seed viability, limits the possibility of applying traditional breeding techniques based on sexual mating and recombination (Sreenath *et al.*, 1994). Nevertheless, Indian researchers have been traditionally involved in the selection of new genotypes of vetiver. Sethi and Gupta (1980) identified 15 new selections superior in oil yield starting from the germplasm of 78 accessions coming from various parts of India.

During the past ten years, vetiver grass has gained renewed attention from scientists and governmental institutions in many parts of the world for its technological application to soil conservation against erosion (Mucciarelli *et al.*, 1997; see also Chapter 6). Its extensive root system grows straight down without interfering with the growth of neighbouring crops and anchors the plant firmly to the ground. Therefore the plant is a powerful environmental tool and it is currently employed by USA and EU governments in bioremediation of eroded lands and in water and soil conservation (Mucciarelli *et al.*, 1997).

Vetiver plants commonly grown outside Asia for its essential oil or for soil conservation have been vegetatively propagated by mean of cuttings (clumps of rootstock). Deliberately spreading an exotic plant around the world may represent a serious danger for natural ecosystems, so the narrow asexual reproduction of "hedgerow" vetiver reduces to a great extent the possible plant escape from culture with uncontrolled spreading like a weed. On the other hand, this infers a very low level of genetic plasticity in cultivated plant populations, leading to the widespread cultivation of a single clone. This could result in progressive erosion of vetiver germplasm with respect to pest resistance and to plant vigour, oil yield and oil quality.

The success of any crop improvement program depends on the extent of genetic variability in the base population; however the depletion of gene pools caused by environmental pressure and human activity leads to a shrinkage of the genetic resources. In this regard cell and tissue cultures are a powerful source of genetic variability and offer to the scientist adequate cell model systems for genetic improvement of most plant crops.

Genetic Variability in Vetiver

Genetic variability in vetiver was initially studied by Kresovich *et al.* (1994). These authors investigated some selected accessions from the United States, employing molecular methods e.g. random amplified polymorphic DNA (RAPD technique) together with rigorous biometric analysis. Their data showed the feasibility and high degree of resolution of the method employed. RAPD patterns were very stable within clones. Non-fertile Huffmann and Boucard cultivars yielded essentially the same genotype, while three samples of the USDA PI 196257 accession (a seed introduction from North India) though morphologically similar were genetically distinct.

With regard to the possible clonal nature of cultivated vetiver grass, Adams and colleagues (1998) applied the same molecular techniques as Kresovich to examine several accessions of *V. zizanioides* coming from various parts of the world. They found that only one genotype, called "Sunshine", accounts for almost all the germplasm utilized outside of South Asia and essentially all of the vetiver used today for erosion control seems to derive from the single Sunshine clone (Adams *et al.*, 1998). On the other hand, when examining a second series of accessions, they found additional variations not only in the Khus vetiver Type of North India, but in other distinct genetic clusters too (Adams and Dafforn, 1997b; Adams *et al.*, 1998). Based on RAPD data, analysis of *V. zizanioides* and other related *Vetiveria* species, *Chrysopogon fulvus* (Spreng.) Chiov. and *C. gryllus* (L.) Trin. revealed that *Vetiveria* and *Chrysopogon* are not distinguishable, suggesting the need for taxonomic revision of these two genera.

All these considerations imply that much genetic diversity of vetiver still awaits discovery. This should prompt us to search for additional non-fertile germplasm to broaden the genetic base of this species in order to diversify the future plantings in erosion control projects.

Vetiver varietal selection through the biomolecular screening of germplasm diversity is promising, as demonstrated by the work conducted at the Central Institute of Medicinal and Aromatic Plants (CIMAP) of Lucknow, India. Starting from biometric data collected from 45 available germplasm accessions maintained at CIMAP, and quantified by multivariate analyses on the basis of plant height, tillers/plant, fresh and dry root yield, essential oil yield etc., sixteen new clones which were superior for high root yield, oil content and oil yield per unit area were selected (Lal *et al.*, 1997a, 1997b). Three new cultivars have now been released for commercial cultivation viz. cv. Dharini, Gulabi and Khesari, with Kush, rose and saffron notes in their oils, respectively (Lal *et al.*, 1998). Clones to be used as prospective parents in hybridization were identified and subsequently characterized on the basis of DNA polymorphism divergence (Shasany *et al.*, 1998), providing scope for future genetic improvement using a marker-assisted approach.

Sexual hybridization imposes narrow limits to the scope of breeding, confining the breeder to the very low variability present in the normal gene pool. Gene transfer to plants via *in vitro* cultures, with the aim of adding specific new genomic traits, makes possible the overcoming of existing problems encountered with sexual hybridization. Genetic transformation of plants involves the stable introduction of foreign DNA sequences into the nuclear or organelle genome of cells capable of giving rise to a whole transformed plant. Circumventing the sexual process and avoiding the need of lengthy backcrossing procedures, genetic transformation may accelerate plant genetic improvement (Draper and Scott, 1991).

In Vitro Biotechnologies

Tissue cultures and micropropagation

The present development of *in vitro* plant propagation and breeding provides a powerful means for the genetic improvement of vetiver grass. Plant propagation using shoot cultures i.e. micropropagation, supplies growers with a highly uniform plant population useful for commercial exploitation and for the maintenance of secure genetic stocks of plant material for breeding, genetic transformation and germplasm conservation. The latter subject is essential for crops which produce short-lived seeds and for those which are normally propagated vegetatively (Withers, 1985) like vetiver grass.

To reach these aims sterile plant material capable of rapid regeneration into mature plants is essential.

Explanting and micropropagation of vetiver

Different plant parts like leaf, rhizome, axillary buds and inflorescences have been used as explant material for tissue and organ culture of vetiver. See Sreenath *et al.* (1994) for a full literature survey.

A very simple procedure for shoot culture and propagation of vetiver is given as follows:

- Young growing sprouts are dissected from plants growing in the field and thoroughly washed in running tap water, external two-three pairs of leaf sheaths being carefully removed and discarded.
- Remaining shoots are incubated for no more than 1–2 days, at room temperature, on wet paper, in order to favour mould growth and spore germination.
- After the shoots have been washed and sterilized with ethanol 70% (3–5 min) the next pairs of leaves are removed.

Table 7.1 Selected Examples of Donor Material and Explanting Procedures of *Vetiveria zizanioides*.

Source of explant	Dissection	Sterilization	Scope	Reference
Rhizomes with axillary buds	Discard leaf bases, scales and old roots	Repeated washings HgCl$_2$ 0.1% (15 min.)	Multiple shooting	Jagadishchandra and Sreenath, 1987
Buds on rhizomes	Discard scale, senescent and upper part of leaves	Repeated washings HgCl$_2$ 0.1% (10–15 min.)	Shoot elongation	Sreenath *et al.*, 1994
Inflorescences and nodal segments	Cut 8–10 cm long culm segments	Repeated washings HgCl$_2$ 0.1% (10–15 min.)	Shooting/ Cell cultures	Sreenath and Jagadishchandra, 1990
Leaf blades, sheath bases and leaf discs	Cut 3–5 mm long mature leaf segments	HgCl$_2$ 0.1% (2–4 min.)	Shooting/ Cell cultures	Mathur *et al.*, 1989

- Complete sterilization is attained by soaking shoots in NaOCl 20% plus Tween 20 or other detergent (0.1%) for 15 min. For older and denser culms, the outermost sheathing leaf pairs can be still discarded and sterilization repeated before washing the shoots with sterile distilled water.
- Shoots are cut at the base and incubated on full strength MS medium (Murashige and Skoog, 1962) supplemented with 3.0 mg l^{-1} 6-benzylaminopurine (BAP) for 2–6 weeks in order to obtain clumping and sprouting (Figure 7.1a).
- *In vitro* grown clumps are then transferred to an MS medium supplemented with 1.5 mg l^{-1} BAP and 0.1 mg l^{-1} indolebutyrric acid (IBA) in order to increase clumping and sprouting (Figure 7.1b).
- Rooting can be enhanced by incubation on MS with 0.1 mg l^{-1} indoleacetic acid (IAA).
- Rooted shoots are separated from clumps, washed with water to remove agar and then treated with a fungicidal solution before acclimatization in a greenhouse. To prevent infections and dehydration, plantlets must be placed in half bottles filled with a 1:1 compost:sand sterile mixture and top wrapped with a transparent plastic seal.

Many other protocols for *in vitro* culture have been established, depending on local needs, cultivars and the aims of the propagation.

Horticultural Systems Inc., one of the seven environmental companies belonging to the Ecogroup International Co., Parrish, Florida, is at present upgrading its micropropagation laboratories to produce over 800,000 plant units per year and vetiver is included in that number. If vetiver grass is further identified and proved to be able to contribute to environmental preservation and restoration on a world-wide scale, the marketing of vetiver will deal with global and economic-social aspects, enforcing the need for much more research in the field of cultivar improvement, bioengineering and genetic engineering.

Figure 7.1 Micropropagated vetiver plants. *a* Plantlets after sprouting and rooting. *b* A single well rooted and clumped vetiver plant, before *ex vitro* acclimatization.

Callus culture

Since Skoog and Miller (1957) earliest observations on hormonal and nutrient culture conditions, it is well assessed that totipotent plant cells can be opportunely stimulated to proliferate producing cell masses known as callus and to regenerate into whole plants via organogenesis (shoot formation) or embryogenesis (somatic embryos formation). Plant cells actively growing in sterile conditions represent unique tools for research studies and applications in plant biochemistry, physiology and genetics. Explants employed for plant transformations with viral or microbial vectors e.g. *Agrobacterium tumefaciens* (Chan *et al.*, 1993; Hiei *et al.*, 1994) vary from individual cells (protoplasts), suspension cultured-cells, callus cells, thin cell layers or tissue slices to whole organ sections (leaves, roots, stems and floral tissues). Therefore plant regeneration starting from cultured material is a fundamental prerequisite of transformation and according to the different target cell several protocols for plant recovery have been developed.

The morphogenetic processes through which plant regeneration is achieved occur directly on *in vitro* cultured plant tissues or indirectly via the production of callus masses.

Callus induction and culture in vetiver

MS medium is commonly employed for the establishment of vetiver callus cultures starting from different tissue sources e.g. leaves, internodes, inflorescences and rhizomes.

Table 7.2 Possible Mechanisms for Producing Transformed Plants Through *In Vitro* Cultures.

Target Plant Material	*Methods of Recovery*
1- Cultured cells	Plant regeneration via a callus phase
2- Immature embryos	Plant regeneration from transformed cell lines through continued development of the embryo
3- Cells in shoot or flower meristems	Chimeric plant production following continued development of meristems. Normal fertilization with pollen derived by transformed inflorescences
4- Pollen	Direct production of transformed plants via fertilization with *in vitro* transformed mature pollen grains
5- Mature or immature zygotes	*In vitro* development of transformed plants
6- Mature plant tissues	Transformed plants production via micropropagation or direct *in vitro* organogenesis

Leaf – Leaf sheath base, taken from primary meristems of the shoot represents a convenient tissue source of explant, as the other hardened and silicized parts of the leaf are unsuitable to induce callus proliferation. In vetiver tissue cultures as for many other graminaceous plants and for cereals best plating efficiencies and callus proliferation are achieved after incubation of explant with the synthetic auxin 2,4-dichlorophenoxyacetic acid (2,4-D) (Rao *et al.*, 1988). Mathur *et al.* (1989) obtained callus proliferation from leaf bases employing 2,4-D (1–2 mg l^{-1}) and kinetin (0.25–1 mg l^{-1}). These authors found that the addition of 40 mg l^{-1} ascorbic acid into the medium helps to overcome tissue and media browning, a very common occurrence in vetiver leaf tissue culture. Calluses grown on these media and subcultured every 8 weeks were pale, white nodular cell masses (Mathur *et al.*, 1989). Similarly to Mathur and colleagues (1989), we obtained a 40% plating efficiency of callus induction, starting from basal low differentiated leaves of vetiver plants of Ethiopian origin (Mucciarelli *et al.*, 1993). Vetiver leaves can be properly sterilized employing 5–10 mm shoot cuttings washed with water, surface-disinfected in 70% ethanol for 10 sec., soaked for 15 min. in a 0.1% solution of sodium hypochlorite plus 2 drops of Tween-20 per 500 ml and finally rinsed several times with sterile distilled water. Leaf explants, consisting of 2–3 mm long slices of rolled leaves were then put in ventilated petri dishes containing 20 ml of MS medium supplemented with 2 mg l^{-1} 2,4-D, 1 mg l^{-1} IAA and 1 mg l^{-1} kinetin (medium A). The response of the leaf explant to the incubation on medium A was the proliferation of a white, translucent callus on the surface of the uppermost cut end (Figure 7.2a and 7.2b). Within 2 weeks from explanting, callus completely covered the leaf tissues and callus masses were transferred to the maintaining medium B, containing 0.2 mg l^{-1} 2,4-D and 0.5 mg l^{-1} kinetin (Mucciarelli *et al.*, 1993) (Figure 7.2b). In medium B cell masses increased their fresh weight with a lag phase of 150 days followed by a log phase during about 80 days with a growth rate of 32 mg fresh weight/day; however calluses on medium B were commonly subcultured every 4 weeks.

In vitro micropropagation offers the advantage of explanting from young and homogeneous plant material, which does not need sterilization and it is in general highly responsive to *in vitro* incubation and especially to plant growth regulator applications.

Figure 7.2 Callus induction and plant regeneration from leaf tissues of *Vetiveria zizanioides*. *a* Proliferation of a white callus on basal leaf explant (bar = 2 mm); *b* Brownish and friable callus subcultured on maintenance medium (bar = 5 mm); *c* Somatic embryos on callus grown on regenerating medium (bar = 500 μm); *d* Portion of callus with shoots regenerating from the embryo-like structures (bar = 5 mm); *e* Well-developed plantlet on rooting medium; *f* Three-month-old regenerated plant successfully established in pot (Mucciarelli *et al.*, 1993).

Optimal callogenesis efficiency was obtained by plating crown and leaf slices taken from *in vitro* micropropagated vetiver plant of Javanese origin (Leupin *et al.*, in preparation). A relatively low concentration of 2,4-D (0.5 mg l^{-1}) has been found more suitable for compact callus induction. Plating efficiency and growth could be increased by adding 0.5 mg l^{-1} BA and increasing MS sucrose concentration up to 75 g l^{-1} (Leupin *et al.*, in preparation).

Inflorescence and nodal segments – Callus formation in vetiver has also been induced in vegetative and floral parts of the culms in the presence of 2,4-D. Once again young immature tissues of rapidly growing stems, floral rachises and spikelet primordia were the only tissues responsive to incubation on culture media (Sreenath *et al.*, 1994). Small pieces of primary callus masses have then been isolated from explant tissues and subcultured every 6–8 weeks on media supplemented with 2,4-D and BAP or kinetin (Sreenath and Jagadishchandra, 1990).

Rhizomes – Owing to the presence of secretory tissues in vetiver roots, the development of protocols for callus induction and culture, starting from root material is also worthy of note. Root-derived cell cultures can be assessed for their *in vitro* essential oil production, that is if these dedifferentiated cultures can retain their biosynthetic capacity. Underground plant organs are normally the worst source of explant owing to the high degree of contamination by bacteria and especially by fungi. Sreenath *et al.* (1994) reported on the response to *in vitro* culture of vetiver rhizomes dissected from in field growing plants. Despite surface sterilization with HgCl$_2$ which was conducted for up to 15 min., more than 90% of the plated rhizomes were contaminated by soil born fungi. Therefore the establishment of *in vitro* well rooted plantlets to be employed as donor material should be taken into account to solve this problem. Mature entire rhizomes responded better than the immature ones and induction of callus took place from a different part of the explant. Here again 2,4-D (1–5 mg l^{-1}) proved to be the best auxin. Roots, axillary buds and the upper cut ends of the rhizomes produced callus. Callus emerging from axillary buds and lateral roots was fast growing and could be easily subcultured. The presence of axillary buds along the surface of the rhizome was essential for callus induction, probably depending on the endogenous hormonal level (Sreenath *et al.*, 1994).

Plant regeneration through embryogenesis

Somatic embryos are thought to arise from single competent cells, activated by a proper nutritional and hormonal treatment. Embryogenic single cells or cell clusters can be easily identified in the culture medium, isolated from the growing substrate and opportunely manipulated. Embryogenic or proembryogenic cells are therefore accessible for gene transfer, rendering genetic transformation possible, and giving rise directly to a non-chimeric transformed plant (Draper and Scott, 1991). This is the reason why plant regeneration through embryogenesis is normally preferred to other more conventional methods of plant recovery including shooting of callus cultures (regeneration via organogenesis).

Proembryogenic cell aggregates develop on primary callus after prolonged culture (5–9 weeks) according to the callus types present. Normally two callus types can be observed and easily identified by accurate observation of the culture plates: a whitish brown, soft or eventually watery callus type, which retains this morphology in the subsequent subcultures and does not become embryogenic, and a pale yellow or whitish, finely nodular, hard callus-type, which become embryogenic and develops many somatic embryos after one to two subculture cycles.

Embryogenesis and plant regeneration in vetiver

Sreenath and Jagadishchandra (1989) reported that supplementing original 2,4-D containing MS medium with BAP or kinetin enhanced the percentage yield of embryogenic callus. The latter originated on the primary leaf-derived callus as nodular compact masses after 30–45 days of culture. Inflorescence-derived callus has been also reported to possess a high embryogenic potential and to retain the proembryogenic capacity even after eight to ten subcultures on the 2,4-D supplemented medium. They proliferated and produced somatic embryos without germinating (Sreenatha and Jagadishchandra, 1990). Callus derived from nodal explants occasionally produced a few roots with numerous root hairs through organogenesis, but never become embryogenic. Calluses produced from different parts of rhizomes were frequently soft and non nodular and have been described as translucent, white, gelatinous and consisting of large vacuolated cells. Rhizome-derived callus failed to show any kind of morphogenesis (Sreenath *et al.*, 1994).

Plants can be easily regenerated starting from embryogenic cultures, when nodular callus, bearing numerous somatic embryos is transferred to the MS basal medium with null or reduced auxin (2,4-D or IAA) concentrations.

Somatic embryos have been described as whitish compact cell structures, first ovoid in shape, the globular stage, and then polarized, and characterized by a typical bipolar axis linked to reserve tissues (*scutellum*) (Emons and De Does, 1993; Emons, 1994). These structures are linked to the original callus through a sort of stalk made of small cells connected directly to the callus tissue. Further embryo maturation consists of the cotyledonary development and the acquisition of the two apical meristems, the shoot and the root primordia.

In vetiver, somatic embryos possess some of these features and became green by 10–15 days, germinating into plantlets. Sreenath *et al.* (1994) reported a maximum of 25–30 plantlets formed from each callus in 45–60 days after transfer to basal MS medium. By about 60 days many regenerated plantlets possessed five to six leaves 5–6 cm long. At this stage plants with a well developed root system were carefully transferred to small pots and kept under greenhouse conditions for further acclimatization. When for plantlets dehydration and pest attack were avoided, a 95–100% survival of plants in field was achieved (Sreenath and Jagadishchandra, 1989, 1990).

Working with crown- and leaf-derived callus from micropropagated plants, we obtained plant regeneration on a modified MS medium containing 0.1 mg l^{-1} 2,4-D, 1.0 mg l^{-1} BAP and 25 g l^{-1} sucrose after an induction period of 6 weeks (Leupin *et al.*, in preparation) (Figure 7.3).

The effect on plant regeneration of higher concentrations of sucrose (10, 15 and 20%) in the MS basal medium had already been tested by Sreenath *et al.* (1994), who found very good plant regeneration at a 10% sucrose level. High sugar concentrations in the medium promoted the maturation of embryos, preventing precocious germination. As documented for *Zea* (Emons and Kieft, 1993) where zygotic embryos normally possess abundant starch reserves, somatic embryos maturation in vetiver seems also to depend on the formation of the scutellar structure. This process takes place only during incubation of cells in media with sugar levels high enough to support

Figure 7.3 Vetiver shoots regenerating from callus cultures. *a* A single shoot apex
emerging from callus (bar = 2 mm); *b* Multiple shooting from vetiver
callus cultures (bar = 5 mm) (Leupin *et al.*, in preparation).

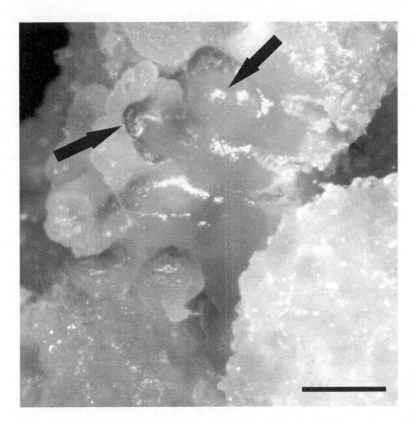

Figure 7.4 Vetiver somatic embryos developing into green shoot apex on the surface of callus. Note smooth, translucent and greening embryogenic surfaces of callus (arrows) (bar = 500 µm).

starch synthesis and accumulation in extraembryonal tissues. Like previous authors, we obtained best performances in plant regeneration when 14-day-old callus was transferred from the original medium for callus induction (medium A) to medium C, which was characterized by an increased concentration of kinetin (2.0 mg l^{-1}) and low levels of 2,4-D (0.2 mg l^{-1}) (Mucciarelli *et al.*, 1993). One month after the transfer to this medium, typical somatic embryos developed on the surface of the callus (Figure 7.2c and Figure 7.4). After 60 days of culture on medium C, 12% of the embryos germinated by producing green shoots with an average of 19 ± 1.8 plants per 160 plated calluses and with a 100% of reproducibility (Figure 7.2d) (Mucciarelli *et al.*, 1993). Embryogenic callus was obtained also on the maintenance medium B with a 10% frequency, but in this medium time necessary to form embryos and subsequent germination was substantially longer than in medium C. Plantlets were then transferred to an MS basal medium without growth regulators (medium D) or to a rooting medium (medium E) containing IAA 0.1 mg l^{-1} to promote root development and complete maturation of vetiver plants (Figure 7.2e). Fifty days later plants were transferred from the aseptic conditions to the outside environment. Regenerated plants were all successfully grown when established in soil (Figure 7.2f) (Mucciarelli *et al.*,

Table 7.3 Cultural and Developmental Stages of Plant Regeneration in *Vetiveria zizanioides*. All Cultures Were Grown on MS Basal Medium Plus 100 mg l^{-1} Casein Hydrolysate and Kept in a Growth Chamber (26/28 °C Day, 21 °C Night, with a 14-h Photoperiod) (Mucciarelli *et al.*, 1993).

Culture medium	2.4-D	IAA	Kin	Culture procedure	Culture target	Incubation period
Medium A	2.0	1.0	1.0	2–3 mm slices of inner leaf	Callus induction	2–3 weeks
Medium B	0.2	–	0.5	Callus selection and subculture. Leaf tissues discard	Callus proliferation and proembryogenic determination	Subcultures every 4 weeks for maintenance
Medium C	0.2	–	2.0	Selection of nodular embryogenic callus	Embryo development and plant regeneration	8–9 weeks
Medium D (E)	–	(0.1)	–	Shoot excision and transfer	Shoot sprouting and plantlet recovery (rooting)	1–2 months

1993). The whole culture procedure time up to complete plant recovery was 5–6 months, as summarized in the table above.

Despite the low value of plant regeneration via organogenesis of callus, it is worth noting the production by Mathur and colleagues (1989) of shoot buds from leaf calluses on a medium containing 1 mg l^{-1} IAA and 0.5 mg l^{-1} kinetin. They reported an 80–95% survival rate for plantlets in the greenhouse and later 96% survival in the field.

Direct plant regeneration

The appearance of genetic instability or somaclonal variation of the cultured material due to the occurrence of chromosome or genetic changes in the regenerated plants is a phenomenon frequently encountered in long-term cultured cells (Evans and Sharp, 1986). Usually these genetic changes are undesirable, especially when clonal reproduction of transformed plants is the goal of the regeneration. In this case, as the occurrence of somaclonal variations increases with the duration and extent of the disorganized or undifferentiated culture phase (Karp, 1994), plants regenerated directly from tissues or organs, eventually transformed with physical DNA delivery methods, are preferred. These techniques allow manipulation of entire vegetative plant organs e.g. leaves and organ meristems or reproductive structures i.e. gametes and zygotes in order to avoid disturbance of normal plant development and reduce the *in vitro* culture phase (Draper and Scott, 1991). Although somaclonal variation is limited to some extent, chimera formation cannot be excluded.

Direct plant regeneration from cultured explants is a method particularly suited for herbaceous dicotyledons, using leaves, corms, bulbs, stems, rhizomes and tubers (Tisserat, 1985). Starting from cultured apical meristems, a cytokinin is usually added to the medium to overcome apical dominance on shooting and to enhance the branching of lateral buds. This method has been successfully applied to the commercial and large-scale propagation of some important ornamental plants.

Direct regeneration in vetiver

Jagadishchandra and Sreenath (1987) reported direct sprouting of the tiny axillary buds of rhizomes into green shoots. Media containing kinetin or BAP promoted sprouting of the axillary buds into shoots. BAP (1 mg l^{-1}) triggered multiple shooting of buds. Buds clumped repeatedly resulting in a bunch of small shoots. Supplying BAP along with IAA (1 mg l^{-1}) the axillary buds developed into individual shoots, amenable for plant propagation. When multiple shooting is unnecessary, vetiver sterile rhizomes can be cultured on MS basal media without growth regulators; their axillary buds become active within 2–3 days and develop into normal shoots of five to six leaves after 15–30 days. The use of basal medium, without any growth regulators, followed by a transfer to a second medium containing IAA (1 mg l^{-1}) allows optimal rooting for the purpose of transplanting to soil (Sreenath *et al.*, 1994).

Sreenath and Jagadishchandra (1990) obtained direct shoot formation starting from both floral primordia and node cultures in the presence of NAA and BAP (1 mg l^{-1} each). Shoots thus obtained can also be rooted on basal medium or in the presence of IAA (1 mg l^{-1}) and established in soil.

At present no information on direct plant regeneration from leaf tissues is available. Regeneration from vetiver leaves could be advantageous considering the growing interest for direct DNA delivery methods for plant transformation. Several techniques are being used today to deliver DNA to isolated plant tissues. This is the case of many important cereals, which among the Poaceae (Gramineae), are the most recalcitrant to cell culture and *in vitro* plant regeneration. Cells which are still surrounded by a cell wall, i.e. cells from suspension cultures or cells in intact leaf tissues require a treatment that can overcome this barrier. At present a method of choice that could be successful in vetiver grass is the delivery of DNA bound to microcarriers that are accelerated by different physical devices e.g. particle bombardment, so that they can enter through the cell wall. Some of these methods have been successfully used and are already commercialized (Fütterer and Potrykus, 1995). Treatments of protoplasts with polyethylene glycol (PEG) (Datta *et al.*, 1990), electroporation (Klöti *et al.*, 1993) or macro- and micro-injection (Lusardi *et al.*, 1994) are now also commonly adopted.

When clonal propagation and genetic stability of cultures are not the main goals, somaclonal variation arising spontaneously from *in vitro* cell cultures is a powerful tool for producing novel and useful varieties. To date as a result of research on somaclonal variants several valuable breeding lines have been developed (Reisch, 1983; Karp, 1994). Vasil and Vasil (1986) observed that much more variability is found in cell clusters than in regenerated plants, and that during regeneration of shoots some degree of selection occurs with the result that only a fraction of the variability of cultured cells is actually recovered in the regenerated plants (Vasil and Vasil, 1986).

Embryogenic cell suspension cultures

Due to the relation existing between cell dedifferentiation and somaclone frequency (Karp, 1994), and to the fact that with shaking, the cell clumps are smaller than on solid medium and often fall apart thus reducing generation of chimeric plants, regeneration from embryogenic cell suspension or protoplast cultures has been exploited in numerous plant species.

Most suspension cultures are obtained by transfer of friable callus clumps to agitated liquid media, and with regard to cell growth induction and proliferation, herbaceous dicotyledons normally respond promptly. On the other hand, in monocotyledons embryogenesis is a prerequisite for the establishment of actively dividing cell suspension cultures.

In most studies on *Poaceae*, suspension cultures have been initiated from immature embryo-derived callus or via anther/microspore cultures. The key step for establishment is the selection on solid media of the proembryogenic material, according to the callus types present, indicated as Type I and Type II in maize. Normally Type I callus is white in colour, compact and nodular, while Type II is soft, friable and translucent (Vasil and Vasil, 1991). Type I callus is a prerequisite for embryogenic cell suspension initiation and for the induction of high division frequencies, particularly useful when large scale production of cell biomass or viable and active protoplast isolation are the aims of the liquid culture. Liquid cultures initiated starting from Type II callus grow slowly, forming large cell clusters rather than a well dispersed cell suspension. This is not a strict rule, since in other species high embryogenic frequencies have been shown by soft Type II calluses as in the case of *Thevetia peruviana* (Kumar, 1992). In the Poaceae family, and for cereals especially, a number of protocols for the establishment of embryogenic suspension lines have already been developed (Vasil and Vasil, 1991). However, establishment has been successful only for a small number of genotypes within these species, suggesting that embryogenic potential is linked to the genetic background of the donor plant material (Kyozuka *et al.*, 1988). Cell suspensions can differ greatly in their embryogenic capacity according to the developmental stage of the donor plant material, the hormone concentration in the media and the growth conditions employed (Funatsuki and Kihara, 1994). From studies conducted directly on cell suspensions it is now clear that embryogenic cells are those cells that are still capable of dividing and of adhering to each other, but have escaped the influence of prolonged auxin treatments, which are essential for cell and tissue dedifferentiation but not for embryogenesis (Emons, 1994). The non-embryogenic cells do not divide; they elongate and only loosely attach to each other or become single cells in the medium. Depending on the time of auxin application, the somatic embryogenic process is direct (short time of auxin application, no subculturing) or indirect (long time of auxin application, frequent subculturing of calluses or proembryogenic masses or PEMs) in liquid media (Emons, 1994). The embryogenic character of a cell suspension cannot be maintained for a long period; suspension cultures of cereals lose their regenerating capacity during prolonged cultures, owing to chromosome loss and rearrangement (i.e. aneuploidy) as a result of the stress generated by the *in vitro* artificial culture system. The loss of cell competence may also be due to the multicellular origin of suspensions, leading to different ploidy levels within the cell population (Krautwig and Lörz, 1995). Several observations have supported the hypothesis that plant growth regulators, auxins especially, are able to induce embryogenesis in culture, alter cell polarity and induce the activation of asymmetric cell division (Dudits *et al.*, 1991).

Cell suspensions establishment in vetiver

As discussed before, the occurrence of somaclonal variations increases with duration of the disorganized phase and the extent of the disorganization, therefore regeneration from a suspension culture would be of great interest in vetiver also. Besides, actively growing cell suspensions of *Vetiveria zizanioides* can be used for studies on vetiver essential oil in *in vitro* biosynthesis and for the establishment of biotransformations protocols.

Table 7.4 Basal Medium for vetiver Callus and Suspension Culture Induction (mg l^{-1}) (Leupin *et al.*, in preparation).

	Modified AA	Modified N6
Macronutrients	AA	N6
Micronutrients	MS	MS
NaFe(III)EDTA	29	37
Thiamine (Vit B1)	10	0.1
Pyridoxine (Vit B6)	1	0.5
Nicotinic acid	1	0.5
myo-Inositol	100	100
Sucrose	20,000	30,000
Sorbitol	25,000	10,000
L-Glutamine	876	200
L-Asparagine	300	113
L-Arginine	174	–
Glycine	7.5	–
L-Proline	–	500

Table 7.5 Liquid media composition (mg l^{-1}).

	AAF	mN6	mN6+B	mN60.5D	mN60.5D0.5B	mN60.1D
	Modified AA			Modified N6		
2,4-D	1	1	1	0.5	0.5	0.1
BAP	–	–	0.5	–	0.5	–
pH :	5.8	5.8	5.8	5.8	5.8	5.8

AA   : (Müller and Grafe, 1878)
N6   : (Chu *et al.*, 1975)
MS   : (Murashige and Skoog, 1962)
2,4-D : 2,4-dichlorophenoxy acetic acid
BAP : benzylaminopurine

Recently, we have approached the study of plant regeneration from embryogenic cell suspensions, starting from compact calluses obtained from non-flowering Javanese vetiver plants (Leupin *et al.*, unpublished results). We studied the influence of original callus from different callus induction media, as well as the influence of different liquid media on the establishment of liquid cultures and on subsequent regeneration of plantlets.

To obtain calluses for liquid culture induction, the *in vitro* plantlets were cut into leaf and crown slices and cultured on different callus induction media. The induction media were composed of modified MS medium supplemented with 2,4-D (0.5 mg l^{-1}), BAP (0 or 0.5 mg l^{-1}) and different concentrations of sucrose (1–10 %). We obtained three types of callus. The gelatinous and the soft callus types were observed first, after 2–4 weeks, whereas a third callus type with compact structures (Figure 7.5a) appeared only after about 4–8 weeks in culture. These calluses with compact structures were subcultured again, before they were used to induce liquid cultures. The calluses were transferred after 3–10 months on callus induction medium into culture vessels containing 15 ml of liquid of each of the following media.

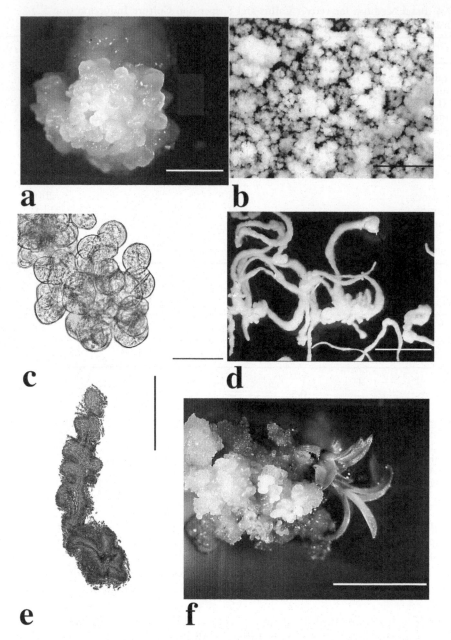

Figure 7.5 Liquid culture induction and plantlet regeneration of *Vetiveria zizanioides*.
a Compact callus as starting material for liquid culture induction (bar = 2
mm); *b* Liquid cultures with compact clumps after about one year on mN6
medium (bar = 5 mm); *c* A typical cell aggregate proliferating in mN6
liquid medium (bar = 50 μm); *d* Root-like structures from mN60.5D0.5B
liquid medium (bar = 5 mm); *e* Section of a root-like structure showing the
absence of tissue differentiation (bar = 2 mm); *f* Regenerated plantlets from
the liquid culture on mN60.5D, after transfer to solid medium (bar = 3 mm)
(Leupin *et al.*, unpublished results).

Cultures were kept at 23 °C in the dark on a rotatory shaker at 70 rpm and were subcultured every second week by replacing the old medium with fresh medium.

For callus induction and also in the liquid cultures it was possible to distinguish three different culture types viz. mucilaginous and thick suspensions, cultures with loose calluses; and suspensions containing compact cell clumps (Figure 7.5b and 7.5c) (Leupin *et al.*, unpublished results). In mucilaginous cultures the medium became viscous and embedded the loose calluses. As good growing liquid cultures were of interest, the mucilaginous cultures were discarded when they remained mucilaginous after another month of subculturing. It took a long time for the cultures to start to form compact cell clumps. Once formed, cultures with compact clumps grew faster than the other cultures, with a doubling time of the fresh weight of about 1 to 2 months.

With respect to starting callus cultures and their influence on subsequent liquid cultures, we found that when calluses were induced on media with 0.5 mg l^{-1} 2,4-D and 10, 25, 50 or 75 g l^{-1} sucrose, no liquid cultures with compact structures were induced subsequently. By adding 0.5 mg l^{-1} BAP to the callus induction medium and increasing the sucrose concentration, more slices with calluses with compact structures were induced (Leupin *et al.*, unpublished results). These changes of the callus induction medium also showed an effect on the induction of liquid cultures with compact clumps. Compact calluses from callus induction medium supplemented with 0.5 mg l^{-1} 2,4-D, 0.5 mg l^{-1} BAP (DB) and low concentrations of sucrose (10 or 30 g l^{-1}/DB10 or DB30) did not give rise to any compact structures in liquid cultures. On the other hand calluses with compact structures from callus induction media with higher concentrations of sucrose (50, 75 or 100 g l^{-1}/DB50, DB75 or DB100) did induce liquid cultures with compact calluses, almost in the same percentages (see following table). This observation could be explained by the fact that with increasing sucrose concentration not only more slices had calluses with compact structures, but also more compact structures per callus were found. Therefore the amount of compact structures increased compared to the amount of soft callus, thus improving the number of compact structures in liquid cultures as well.

The following table describes results concerning the influence of callus induction and liquid media composition on suspension establishment in vetiver.

The difference in the induction of liquid cultures with the modified N6 or the modified AA basal medium was minor. However, since on mN6 somewhat more cultures with compact clumps were induced than on AAF, we continued with the modified N6 basal medium.

As the growth regulator BAP in the callus induction medium had a beneficial effect on the induction of calluses with compact structures and on the subsequent induction of liquid cultures, 0.5 mg l^{-1} BAP was added to the liquid culture medium. Unfortunately, the additional BAP had not the same beneficial effect in liquid medium. It influenced neither the induction of compact structures nor the induction of the other liquid culture types.

In earlier experiments with callus induction media higher 2,4-D concentrations induced more gelatinous calluses (Leupin *et al.*, in preparation). Therefore, to reduce the development of the mucilaginous liquid cultures and maybe to obtain more cultures with compact structures, the 2,4-D concentration was lowered from 1 to 0.5 or even to 0.1 mg l^{-1}. This resulted in fewer mucilaginous cultures, but the percentage of liquid cultures with compact clumps did not increase. For the addition of 0.1 mg l^{-1}

Table 7.6 Morphological Features and Time Necessary for the Induction of Cell Suspensions in Vetiver With Respect to the Composition of Medium Employed for Starting Both Callus and Liquid Cultures (Leupin et al., unpublished results).

Callus Induction		Liquid Medium[b]	Culture (numbers)	Induction time (months)	Development of the Culture on Liquid Medium		
Medium[a]	Time (months)				Compact cultures %	Loose cultures %	Mucilaginous cultures %
DB10	4	mN6	10	10	0	20	80
DB30	4	mN6	8	10	0	0	100
DB50	4	mN6	11	10	18	0	82
DB75	3–5	mN6	124	6–14	11	13	76
DB100	8–10	mN6	24	7–8	13	*	*
DB75	4–5	AAF	38	8–14	5	11	84
DB75	3–5	mN6+B	62	8–14	10	14	76
DB75	3	mN60.1D	13	6	0	100	0
DB75	3	mN60.5D	25	9	8	–	**
DB75	3	mN60.5DOB	25	9	0	–	**

a: DBx: modified MS medium supplemented with 0.5 mg l^{-1} 2,4-D, 0.5 mg l^{-1} BAP and x g l^{-1} sucrose.
b: Composition of the liquid media as previously reported.
*: No data.
**: There were many more loose cultures than mucilaginous cultures.

2,4-D no cultures with compact clumps were induced and the cultures turned brown. The reduction of the 2,4-D concentration in the medium also had another effect on the cultures: in some cultures root-like structures were found (Figure 7.5d). Some of these structures even produced something like side-roots. Histological observation showed that these structures were not roots, but callus growing in a root-like fashion (Figure 7.5e). Addition of 0.5 mg l^{-1} BAP to the mN60.5D medium resulted in more cultures with root-like structures and no compact clumps.

Plant regeneration from suspension cultures of vetiver

For the regeneration experiments compact clumps from the liquid cultures were transferred to solid callus induction media DB75 or DB10 (solid modified MS medium with 0.5 mg l^{-1} 2,4-D, 0.5 mg l^{-1} BAP and 75 or 10 g l^{-1} sucrose). The resulting calluses were subsequently transferred to regeneration media to test the ability of the culture to regenerate plantlets. For regeneration the 2,4-D concentration was either reduced to 0.1 mg l^{-1} and the BAP concentration was increased to 1 mg l^{-1} (D0.1B1) or 2,4-D was omitted and 0.5 mg l^{-1} BAP (B0.5) or 1 mg l^{-1} kinetin and 0.1 mg l^{-1} IAA (VRM8) were added. For all three regeneration media the modified MS medium was supplemented with 25 g l^{-1} sucrose and 0.65% agar (Leupin et al., unpublished).

Some of the clumps became brown, others remained white and grew as fine granular callus, a few produced bigger compact structures which became green and from two cultures we were able to regenerate a few plantlets (Figure 7.5f). Both cultures regenerated on DB75 callus induction medium followed by D0.1B1 regeneration medium. The two calluses which regenerated plantlets came from two different cultures. One callus developed from a 9 months old mN6 liquid culture, the other from a 6 months old mN60.5D liquid culture. At this stage these cultures were a mix between the original calluses, loose calluses and compact clumps. As afterwards no more plantlets were regenerated from these cultures, it is not clear whether the plantlets were regenerated from the liquid culture or simply represented carry-over primordia from the original material. After 14 months, when enough compact clumps in the liquid cultures were available, they did not regenerate any more (Leupin et al., unpublished results). To regenerate more plantlets from the liquid cultures, a method of obtaining faster liquid cultures or prolonging the regeneration ability should be found.

Several factors influence the production of liquid cultures including the starting material, the composition of the liquid medium, the treatment of the cultures and environmental conditions (temperature, shaking speed, cultivation in dark/light, etc.).

The starting material seems to be an important factor, as in the experiments, it was shown that the calluses induced on different callus induction media had an effect on the liquid culture induction. The changes of the liquid media (mN6 or AAF) or different concentrations of growth regulators (1 or 0.5 mg l^{-1} 2,4-D and 0 or 0.5 mg l^{-1} BAP) did not show any effect on the induction of liquid cultures with compact clumps. The improvement of the liquid culture should be started first with the improvement of the starting callus. One possibility would be to change the ratio between soft and compact callus in the starting material by tearing the calluses into small pieces and discarding any soft callus present. Another possibility would be to improve the callus induction medium to obtain a better callus type from which the liquid culture could be induced faster.

One problem in our process was that the callus clumps inoculated remained more or less intact and did not split up. Patnaik *et al.* (1997) were able with palmarosa to establish suspension cultures with cell aggregates by sieving and resuspending the culture in fresh medium. This method has also the advantage that carry-over primordia from the original material will be removed in time.

Nutrient medium composition: studies on sugar uptake in liquid cultures of vetiver

An important factor controlling cell growth and morphogenetic potential of plant cell cultures is the availability and uptake of nutrients. Sugars are the main carbon source for non-photosynthetic cell cultures. It has been demonstrated that sucrose administered externally to plant cell suspensions is usually rapidly hydrolysed to glucose and fructose which are then taken up by passive or active processes, depending on the species (Stepan-Sarkissian and Fowler, 1986).

The understanding of the fine control of sugar uptake in cell suspensions from *V. zizanioides* is useful in the establishment of protocols for good quality biomass production, with consideration also of possible future exploitation of large-scale cultivation in bioreactors.

With respect to energy demand, the culture cycle follows three main phases. In the first phase cell loading of carbon and nitrogen substances occurs with rapid uptake of sugars and nitrogen from the medium. Oxidation is low and the sugars are stored as starch. In the second phase cell division occurs with oxidation of carbon sources and high respiratory rates. In carbon-limited media the carbohydrate reserves are remobilized and generated as carbon skeletons for both oxidation and the synthesis of structural components and secondary metabolites (Cresswell *et al.*, 1989). In many culture systems cell growth in terms of biomass increase and secondary product level were two separate phenomena. Understanding the mechanisms which direct and regulate the timing of cell growth allows determination of the sugar to be employed to attain excellent biomass production and to establish a culture procedure to improve secondary product yield.

At cell and subcellular level undifferentiated cell cultures lack the membrane and organelle specialization necessary for a complete expression of the cell enzymatic apparatus and compartmentalization for secondary metabolism. In cell suspension and callus cultures, glandular hairs as well as other secretory cell structures are absent. This causes an overall energy flow in the direction of primary metabolism, lowering the cell biosynthetic potential for secondary products. The addition of an accumulating liquid or solid phase such as a lipophilic substrate or ion exchange resins can alter not only the proportion of secondary products accumulating in the medium, but also the total cell production. Recently the use of polymeric adsorbents (XAD-resins) in liquid media has proved to be effective in eliciting secondary metabolite production and release from cultured plant cells (Strobel *et al.*, 1991). Their mechanism of action could pass through:–

- The protection of volatile or easily degraded secondary products from oxidation in culture media;
- Reduction of feedback inhibition in cell metabolism and detoxification of nutrient media;

- Cell de-repression or stimulation at the genome level;
- Improvement of cell nutrient uptake.

With regard to liquid cultures, we demonstrated the eliciting effect of Amberlite XAD-4 added to non-photosynthetic liquid culture of *V. zizanioides* (Mucciarelli *et al.*, 1994). The effect of addition to the medium of the non-ionic polymeric adsorbent XAD-4 (Amberlite) was a stimulation of both cell growth and viability. These parameters were constantly higher than in control cultures and reached their maximal values after 15–20 days of incubation (Figure 7.6a). Light microscope observations showed a great tendency towards starch synthesis and storage in the form of grains finely dispersed in the cytoplasm of XAD-4 incubated cells (Figure 7.6b). During incubation sucrose was completely metabolized and its concentration dropped to zero both in XAD-4 treated and control cultures, with the release into the medium of glucose and fructose (Mucciarelli *et al.*, 1994). From these preliminary observations we tried to explain the eliciting effect of Amberlite XAD-4 and its possible action on vetiver uptake of sucrose, glucose and fructose by means of radio-labelled feeding experiments.

Callus suspension cultures were established by transferring 6 g (fresh weight) callus in 80 ml of MS medium containing 2% sucrose (w/v) and modified according to Mucciarelli *et al.* (1993). Suspension cultures were maintained on a gyratory shaker at 25 rpm for 30 days. Amberlite XAD-4, tried as a growth elicitor, has been tested in activated and non-activated form and added sterile at 2% (w/v) in liquid cultures (Camusso *et al.*, 1999; Camusso *et al.*, in preparation). Glucidic compositions of cell culture and suspension media were determined by fractionating extraction and enzymatic hydrolysis and analysis by GC-MS, after resuspension in pyridine and derivatization by 1,1,1,3,3,3-hexamethyldisilazane (HMDS) and trimethylchlorosilane (TMCS).

Uptake experiments were conducted in MES-NaOH buffer pH 5.70 added with K_2SO_4 0.125 mM and $CaSO_4$ 0.5 mM to facilitate sugar absorption. Radio-labelled sugars were added to the incubation solutions at 2, 5, 10, 25 and 50 mM concentrations and incubated for 1, 3, 5, 20 and 30 min. Samples were withdrawn and filtered and the biomass was washed with MES-NaOH buffer and extracted overnight in 80% ethanol at 50°C. Washing solutions were immediately analysed with a Packard 1500 TRI-CARB liquid scintillation analyser with a Pico-Fluor 40 quencher. The specific activities of [UL ^{14}C]-sucrose, [UL ^{14}C]D-glucose and [UL ^{14}C]-fructose were 565 mCi/mmol, 310 mCi/mmol and 289 mCi/mmol respectively (Camusso *et al.*, 1999; Camusso *et al.*, in preparation).

In this second set of experiments XAD-4 treated cultures showed again higher viability (75%) and biomass production with respect to control cultures (Figure 7.7), and we demonstrated that this occurrence depended on an enhanced ability to use glucose and fructose by XAD-4 elicited cells (Camusso *et al.*, 1999; Camusso *et al.*, in preparation).

Both sucrose and glucose assured the growth of liquid cultures of *V. zizanioides*, but cultures grown on sucrose showed a higher mean viability at the end of the experimental trials. However, after a starvation period of 12 hours conducted in $CaSO_4$, the analysis of uptake kinetic for sucrose, glucose and fructose indicated that glucose owned the major affinity with a $K_m = 1,7$ nmol, $V_{max} = 45$ nmol $h^{-1}kg^{-1}$ fresh weight (FW), followed by sucrose $K_m = 3,2$ nmol, $V_{max} = 11$ nmol $h^{-1}kg^{-1}$ FW and fructose, $K_m = 2,4$ nmol, $V_{max} = 6,5$ nmol $h^{-1}kg^{-1}$ FW (Camusso *et al.*, 1999; Camusso *et al.*, in preparation).

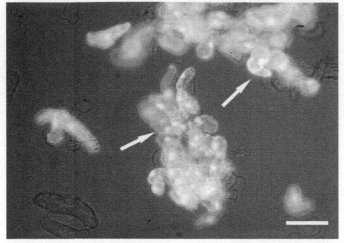

a

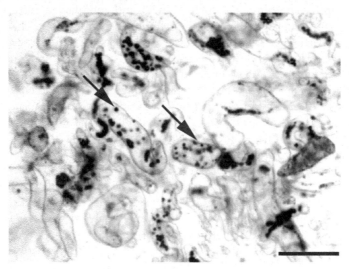

b

Figure 7.6 Vetiver cell suspension in the presence of Amberlite XAD-4 resin. *a* Viable
cells stained with fluorescein diacetate (white arrows) (bar = 100 μm);
b Starch granules in vetiver cell suspensions after iodine treatment (black
arrows) (bar = 50 μm).

In XAD-4 treated cultures we observed a general increase in the affinity for the three
carbon substrates: glucose still possessed the highest affinity (K_m = 4,2 nmol, V_{max} =
1,2 μmol $h^{-1}kg^{-1}$FW), but fructose affinity was higher than sucrose with K_m = 1,8
nmol, V_{max} = 113 nmol $h^{-1}kg^{-1}$FW and K_m = 1,7 nmol, V_{max} = 29 nmol $h^{-1}kg^{-1}$FW,
respectively. Radio-labelled sugar uptake experiments allowed us to indicate the
following affinity order for control and XAD-4 treated liquid callus cultures viz.

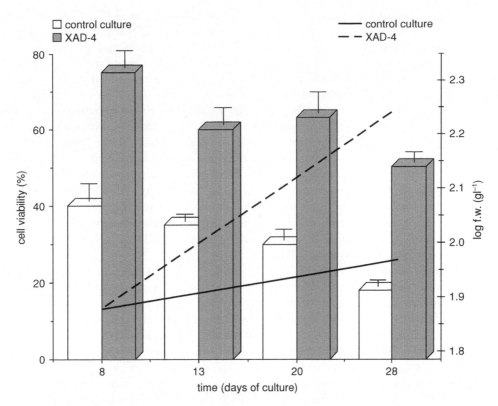

Figure 7.7 Cell viability and cell growth of vetiver suspension cultures with respect to Amberlite XAD-4 elicitation. The addition of XAD-4 to the liquid medium promotes both cell growth (lines) and viability (bars). At the end of the culture period (28th day), 50% of XAD-4 treated cells are still alive and proliferating (Mucciarelli *et al.*, 1994).

glucose> sucrose> fructose and glucose> fructose> sucrose, respectively, treated cultures showing higher K_m and V_{max} than control cultures (Camusso *et al.*, 1999; Camusso *et al.*, in preparation). Therefore we could conclude that the addition of a polymeric adsorbent to liquid cultures of vetiver allowed the optimization of culture parameters, promoting cell growth and viability via a direct effect on vetiver sugar uptake. A detoxifying action of the resin against both chemical impurities of the growing medium and cell released metabolites could therefore be excluded in favour of a more direct action of XAD-4 towards cell primary metabolism. Our study suggested reconsidering the nature of carbon sources to be added to nutritional media for cell cultivation and indicated XAD-4 addition to the media as an effective tool for cell biomass optimization.

Morphological aspects of embryogenesis and plant regeneration

Cytological and histological aspects of cell and callus masses are also important determinants of the embryogenic process. In the *Poaceae*, specific meristematic centres with proembryogenic value have been described as originating from cells belonging to the vascular cambium or surrounding vascular bundles of the explant (Wernicke and Brettell, 1980). The presence of starch granules in the cytoplasm is another common feature of embryogenic cells or tissues cultured *in vitro* (Schwendiman *et al.*, 1988). Embryogenic suspension cultures contain normally a number of different cell types, present both in the PEMs which are cell aggregates consisting of small cytoplasm-rich dividing embryogenic cells and among the single cells dispersed in the liquid medium. In both cases, larger cells with large vacuoles normally become elongated and do not divide while small densely cytoplasmic cells actively proliferate and become embryogenic after auxin removal from the media (Emons, 1994). Both in PEMs of liquid cultures and in embryogenic callus aggregates the small embryogenic cells are connected with the large elongated cells, which have a suspensor-like function, providing nutrients and hormones to non-photosynthesizing somatic embryos. In all these cases the first prerequisite for embryo development is a cell that is able to divide and produces daughter cells which remain attached to each other through a number of cell divisions that are not followed by cell enlargement. Thus a globular multicellular structure is formed (Emons, 1994).

Once developed on the callus surface, somatic embryos have intrinsic polarity and this attachment is essential for complete embryo development and for the formation of a proper shoot meristem. This situation is comparable to the attachment of the zygotic embryo to the mother plant. In *Daucus carota*, which easily forms free complete somatic embryos in liquid cultures, polarity is less pronounced with respect to an embryogenic callus on solid media. However, when pro-embryos are still attached to non-embryogenic cells, polarity is found in the distribution of the calcium-calmodulin complex which concentrates in the future root side of the embryo (Timmers *et al.*, 1989). In species such as *Zea mays*, in which the zygotic embryo has a large storage organ, the scutellum, successful regeneration of somatic embryos depends on the formation of this organ. After an embryo maturation phase in a medium with high sugar concentration, *scutellum*-like tissues are formed, starch accumulates in the cells and embryo polarity development ultimately results in the formation of the two root and shoot meristems (Emons, 1994).

We studied tissue dedifferentiation and callus development starting from nodal explants of *V. zizanioides* after incubation on MS medium supplemented with 2,4-D 2.0 mg l^{-1}, IAA 1.0 mg l^{-1}, kinetin 1.0 mg l^{-1} and modified according to Mucciarelli *et al.* (1993). The last apical node together with the whole shoot apex was excised from plants, sterilized, cut into four segments and incubated on the agar medium. Plant material was treated for conventional optical microscopy, sampling at two days intervals up to the 8th day of incubation and then every eight days of culture up to shoot emergence from callus (Mucciarelli *et al.*, in preparation).

Microscope analyses of plant material revealed that incubation on the medium determined a quick dedifferentiation of explant tissues, starting from the very first days of culture. Tissue dedifferentiation was induced by the presence of the plant growth regulators added to the nutrient medium. In the case of explant material coming from stem portions (the two lowest segments cut from the apical shoot) tissue dedifferentiation consisted in the following steps:–

- Appearance of areas with degenerating cells, right below stem epidermis;
- Lignification and differentiation in tracheids of parenchymatic cells dispersed into the cortical parenchyma or lining stem vascular bundles;

- Active cell proliferation with periclinal and anticlinal divisions of cell aggregates localized under the epidermis of the stem (Figure 7.8a);
- Mitotic activity of provascular tissues leading to *de novo* xylem differentiation in vascular bundles (Figure 7.8b);
- Cortical cells proliferation leading to callus aggregates which protruded outside the epidermis.

Different was the case of leafy portions of cultured explants (the two uppermost segments cut from the shoot) where cell dedifferentiation started from mesophyll, bundle sheath and epidermal cells (Figure 7.8c). Similarly to *Vetiveria*, in *Sorghum bicolor* and in *Saccharum officinarum* cells which responded by dividing to 2,4-D treatment are those surrounding vascular bundles (Wernicke and Brettell, 1980; Ho and Vasil, 1983).

After 8 days from incubation both stem and leafy explants of vetiver developed callus aggregates surrounding the residual epidermis of the explant. After 16 days incubation, callus was clearly visible on the surface of explants (Figure 7.8d). Callus cell aggregates consisted in an inner portion made of polygonal cells with dense cytoplasm and narrow intercellular spaces (Figure 7.8e). Going to the outer cortex of the callus mass, cells become vacuolated and elongated, loosing progressively cell to cell contacts (Figure 7.8e and 7.8f) (Mucciarelli *et al.*, in preparation).

These findings are corroborated by the previous experiments of Sreenath (1983), who described the presence of few clusters of meristematic cells in rhizome calluses derived from a triploid variety of vetiver. The meristematic cells were small with large nuclei and nucleoli and dense cytoplasm, embedded in a cortical layer of large elongated cells. Small cells showed normal mitotic divisions and retained their parental chromosome number up to twelve months of culture (Sreenath, 1983).

On our vetiver callus regenerating shoots appeared after two months of culture. Two kinds of regenerating structures were present (Figure 7.9). A shoot-like type consisting of an apical meristem flanked by leaf primordia and young leaflets (Figure 7.9a) and an embryo-like type corresponding to a somatic embryo (Figure 7.9b). The latter consisted in a structure having intrinsic polarity, a shoot and a root primordia, attached to the original callus through a connecting tissue extending from the root apex to the basal callus (Mucciarelli *et al.*, in preparation).

Evidence of organogenetic processes in embryogenic calluses have also been found in other *Poaceae* e.g. in *Triticum aestivum* (Ozias-Akins and Vasil, 1982) and *Elaeis guineensis* (Schwendiman *et al.*, 1988). Organogenesis as well as xylogenesis in tissue cultures (Gresshoff, 1978) are stimulated by higher level of auxins in the medium with respect to those normally employed to induce embryogenesis. Therefore a low concentration of total auxin in the medium, such as that employed by us more recently (Leupin *et al.*, in preparation) should allow an optimization of callus embryogenic efficiency. In vetiver somatic embryos a true scutellar tissue has never been observed. This occurrence could depend on the low concentration of carbon source used for regenerating media up to date. Higher sucrose concentrations are probably essential for complete maturation of this tissue and sufficient accumulation in it of starch reserves (Emons, 1994). In this regard, MS medium modified by the addition of more sucrose (Leupin *et al.*, in preparation) and/or with the addition of XAD-4 resin (Camusso *et al.*, 1999; Camusso *et al.*, in preparation) should favour the maturation and germination of somatic embryos, giving higher regenerating efficiency in vetiver.

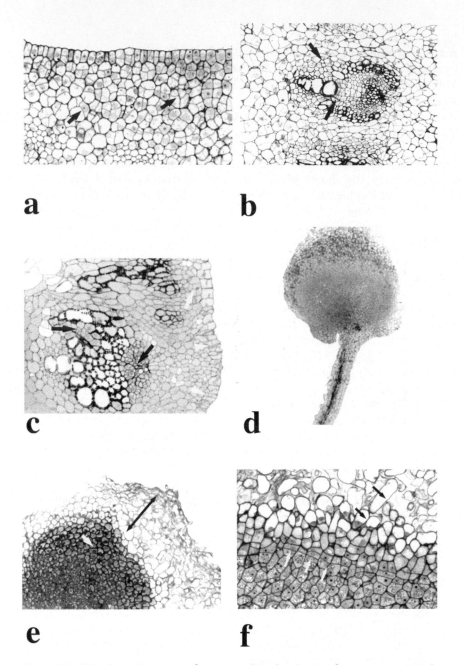

Figure 7.8 Histological aspects of vetiver callus developing from stem and leaf explants. *a* Cell proliferation (arrows) in a 6-day-old stem explant (546x); *b* Mitotic activity and tracheary element differentiation (arrows) in a vascular bundle of a 6-day-old stem explant (336x); *c* Tissue dedifferentiation (black arrows) and mitotic activation of mesophyll, vascular bundle and epidermis cells (white arrows) in a 8-day-old leaf explant (336x); *d* A 16-day-old callus proliferating on the cut end of a vetiver leaf (336x); *e* Cross section of a callus showing an inner portion with small cells (white arrow) and an outer cortex of enlarged cells (black double arrow) (546x); *f* Cross section of a cell aggregate at the boundary between the innermost proliferating cell layers (white arrows) and the outermost vacuolated cells (black arrows) (1080x) (Mucciarelli *et al.*, in preparation).

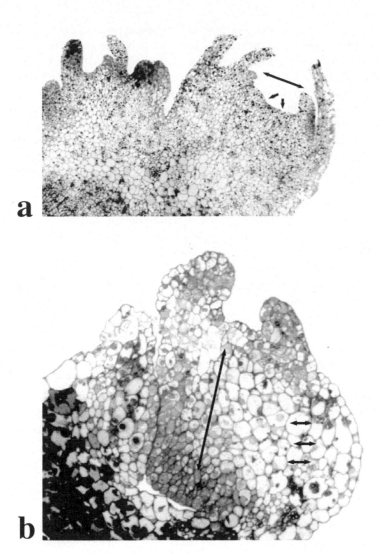

Figure 7.9 Regenerating vetiver callus. *a* Longitudinal section of two shoot apex originating from an organogenic callus. Note the apical meristem (single arrows) and lateral leaf primordia (double arrow) (134x); *b* A longitudinal section of a vetiver somatic embryo at the beginning of formation. Note embryo shoot-root axis (double arrow) and tissue side connections to the basal callus (little double arrows) (546x) (Mucciarelli *et al.*, in preparation).

Concluding Remarks

The establishment of transformation protocols for monocotyledons in general and particularly for cereals and grasses was delayed with respect to transformation protocols for many dicotyledons. Many Poaceae, including all the economically important species, have only been successfully transformed during the last 10 years. Transformation of most of the species however is still unsuitable for genotypes of crop value and must be better refined to reach high efficiencies and commercialization. The reason for this is that the success of a genetic transformation depends on genotype availability, effective selection of transformed plants and on the cell line or cultivar response to *in vitro* culture. Many Poaceae showed both strict genotype dependence and *in vitro* tissue culture recalcitrance and this fact has slowed their biotechnological exploitation so far. In this regard, genetic diversity of vetiver still needs investigation and prompts us to search for new germplasm accessions since transformation technology must be transferred to lines of true agronomical interest.

The majority of genetically transformed plants studied to date have been generated via organogenesis or embryogenesis in tissue cultures with the presence of selective agents. Therefore the host range for transformation is still dependent on the availability of tissue culture techniques.

In this chapter we have focused on vetiver cell and tissue culture methods, and it has been shown that an efficacious protocol for *in vitro* embryogenic cell development and plant regeneration of vetiver is now available. This finding opens the way for other important developments in vetiver cell tissue culture technology. Having defined the right source of explant and culture procedures, protoplast isolation and processing in vetiver is now feasible. Protoplast technology broadens the source of somaclonal variation obtainable from tissue cultures and alternatively is suitable for direct DNA delivery techniques through the mean of microprojectiles, microinjection or electroporation. In addition, the aforementioned embryogenic potential of vetiver cell is a fundamental prerequisite for cell suspension culture establishment that can be exploited for the *in vitro* generation of perfumes and fragrances that are becoming increasingly attractive.

For any biotechnological application it is extremely important to preserve and investigate the basal genetic pool of the species in order to find useful traits that might be improved or engineered. Genetic bases controlling plant fertility and seed viability are of great interest in vetiver biotechnology for two main reasons. First, as many of the physiological and morphological abnormalities exhibited in plants derived from tissue cultures are epigenetic, being transiently expressed only during the *in vitro* phase, it is important to concentrate studies on gene expression in seed populations derived directly from the original transformed plant. Secondly, the general plant sexual fertility and particularly the release and spreading of transgenic pollen into the environment are genetic traits to be controlled when engineered plants are employed for extensive plantings as in the case of erosion control projects.

The number of useful traits that might be engineered into crop plants is growing, and it is greatly stimulated by the increasing identification of new genes controlling important crop characteristics. In this regard, two principle lines challenge biotechnologists working on vetiver, plants with higher protein content in the leaves in order to allow animal feeding with crop wastes and the selection of cold acclimated varieties with the aim of extending vetiver technology to the northern regions of the globe.

Acknowledgements

We are grateful to Prof. B. Witholt (ETH, Zürich), Prof. S. Scannerini (Dept. Plant Biology, Turin), Prof. K.H. Erismann (Phytotech Labor, Bern), Prof. T. Sacco (Veterinary Faculty of Turin), Dr. G. Spangenberg and Dr. Z.Y. Wang (ETH, Zürich), for their advice, encouragement and laboratory facilities. We wish to thank also Dr. W. Camusso (Dept. Plant Biology, Turin) for radio-labelled sugar analysis and Dr. A.

Fusconi (Dept. Plant Biology, Turin) for optical microscopy of vetiver cultures. The vetiver plants from Java were provided by Mr. Heini Lang (Jakarta). The work of R.E.L. was supported by Givaudan-Roure Forschung AG Dübendorf, Switzerland and by the Swiss Federal Office for Economic Policy, project no. 2561.1 of the Commission for Technology and Innovation.

References

Adams, R.P., Zhong, M., Turuspekov, Y., Dafforn, M.R. and Veldkamp, J.F. (1998) DNA fingerprinting reveals clonal nature of Vetiveria zizanioides (L.) Nash, Gramineae and sources of potential new germplasm. *Molecular Ecology*, 7, 813–818.

Adams, R.P. and Dafforn, M.R. (1997a) Lessons in diversity: DNA sampling of the pantropical vetiver grass uncovers genetic uniformity in erosion-control germplasm. *Diversity*, 13(4), 133–146.

Adams, R.P. and Dafforn, M.R. (1997b) DNA fingerprints (RAPDs) of the pantropical grass vetiver, Vetiveria zizanioides (L.) Nash. (Gramineae), reveal a single clone "Sunshine", is widely utilized for erosion control. *Vetiver Newsletter*, 18, 27–33.

Camusso, W., Mucciarelli, M., Sacco, S., Scannerini, S. and Maffei, M. (1999) Sugar uptake in *Vetiveria zizanioides* cultured *in vitro*. 38th *Italian Plant Physiology Congr.*, Turin, September 1999.

Chan, M.T., Chang, H.H., Ho, S.L., Tong, W.F. and Yu, S.M. (1993) *Agrobacterium*-mediated production of transgenic rice plants expressing a chimeric α-amylase promoter/β-glucuronidase gene. *Plant Mol. Biol.*, 12, 505–509.

Chu, C.C., Wang, C.C., Sun, C.S., Hsu, C., Yin, K.C., Chu, C.Y. and Bi, F.Y. (1975) Establishment of an efficient medium for anther culture of rice through comparative experiments on the nitrogen sources. *Scientia Sinic.*, 18, 659–668.

Cresswell, R.C., Fowler, M.W., Stafford, A. and Stepan-Sarkissian, G. (1989) Inputs and outputs: primary substrates and secondary metabolism. In W.G.W. Kurz (ed.), Primary and Secondary Metaboilsm of Plant Cell Cultures II. Springer-Verlag, Heildeberg, pp. 14–26.

Datta, S.K., Peterhans, A., Datta, K. and Potrykus, I. (1990) Genetically engineered fertile indicarice recovered from protoplasts. *Biotechnology*, 8, 736–740.

Draper, J. and Scott, R. (1991) Gene transfer to plants. In D. Grierson (ed.), Plant Genetic Engineering, Blackie, Glasgow, pp. 38–41.

Dudits, D., Bogre, L. and Gyorgyey, J. (1991) Molecular and cellular approaches to the analysis of plant embryo development from somatic cells *in vitro*. *J. Cell Sci.*, 99, 475–484.

Emons, A.M.C. and De Does, H. (1993) Origin and development of embryo and bud primordia during maturation of embriogenic calli of *Zea mays* L. *Can. J. Bot.*, 71, 1349–1356.

Emons, A.M.C. and Kieft, H. (1993) Histological comparison of single somatic embryos of maize from suspension culture with somatic embryos attached to callus cells. *Plant Cell Rep.*, 10, 465–488.

Emons, A.M.C. (1994) Somatic embryogenesis: cell biological aspects. *Acta Bot. Neerl.*, 43(1), 1–14.

Evans, D.A. and Sharp, W.R. (1986) Application of somaclonal variation. *Biotechnology*, 4, 528.

Funatsuki, H. and Kihara, M. (1994) Influence of primary callus induction conditions on establishment of barley cell suspensions yielding regenerable protoplasts. *Plant Cell Rep.*, 13, 551–555.

Fütterer, J. and Potrykus, I. (1995) Transformation of Poaceae and gene expression in transgenic plants. *Agronomie*, 15, 309–319.

Gresshoff, P. (1978) Phytohormones in growth and differentiation of cells and tissues cultured *in vitro*. In P.B. Letham, P.B. Goodwin and T.J.V. Higgins (eds.), *Phytohormones and related compounds – A comprehensive treatise*. Elsevier, New York, vol. 2, pp. 1–29.

Hiei, Y., Ohta, S., Komari, T. and Kumashiro, T. (1994) Efficient transformation of rice (*Oryza sativa* L.) mediated by *Agrobacterium* and sequence analysis of the boundaries of the T-DNA. *Plant J.*, 6, 271–282.

Ho, W. and Vasil, J. (1983) Somatic embryogenesis in sugarcane (Saccharum officinarum L.). I. The morphology and physiology of callus formation and the ontogeny of somatic embryos. *Protoplasma*, 118, 169–180.

Jagadishchandra, K.S. and Sreenath, H.L. (1987) In vitro culture of rhizomes (root stock) in *Cymbopogon* Spreng and *Vetiveria* Bory. In G.M. Reddy (ed.), *Recent advances in plant cell and tissue culture of economically important plants*, Osmania University, Hyderabad, pp. 199–208.

Karp, A. (1994) Origins, causes and uses of variation in plant tissue cultures. In I.K. Vasil and T.A. Thorpe (eds.), *Plant Cell and Tissue Cultures*, Kluwer Academic Publisher, Dordrecht, pp. 139–151.

Klöti, A., Iglesias, V.A., Wünn, J., Burkhardt, P.K., Datta, S.K. and Potrykus, I. (1993) Gene transfer by electroporation into intact scutellum cells of wheat embryos. *Plant Cell Rep.*, 12, 671–675.

Krautwig, B. and Lörz, H. (1995) Cereal protoplasts. *Plant Science*, 111, 1–10.

Kresovich, S., Lamboy, W.F., Li, R., Ren, J., Szewc-McFadden, A.K. and Bliek, S.M. (1994) Application of molecular methods and statistical analyses for discrimination of accessions and clones of vetiver grass. *Crop Science*, 34, 805–809.

Kumar, A. (1992) Somatic embryogenesis and high frequency plantlet regeneration in callus cultures of *Thevetia peruviana*. *Tiss. Org. Cult.*, 31, 47–50.

Kyozuka, J., Otoo, E. and Shimamoto, K. (1988) Plant regeneration from protoplasts of indica rice: genotypic differences in culture response. *Theor. Appl. Genet.*, 76, 887–890.

Lal, R.K., Sharma, J.R. and Misra, H.O. (1997a) Genetic diversity in germplasm of vetiver grass, *Vetiveria zizanioides* (L.) Nash ex Small. *J. Med. Arom. Plant Sc.*, 5(1), 77–84.

Lal, R.K., Sharma, J.R. and Misra, H.O. (1997b) Varietal selection for high root and oil yields in vetiver, *Vetiveria zizanioides*. *J. Med. Arom. Plant Sc.*, 19(2), 419–421.

Lal, R.K., Sharma, J.R., Naqvi, A.A. and Misra, H.O. (1998) Development of new varieties – Dharini, Gulabi and Kesari of vetiver (Vetiveria zizanioides). *J. Med. Arom. Plant Sc.*, 20(4), 1067–1070.

Lusardi, M.C., Neuhaus-Url, G., Potrykus, I. and Neuhas, G. (1994) An approach to genetically engineered cell fate mapping in maize using the Lc gene as a visible marker; transactivation capacity of Lc vectors into somatic embryos and shoot apical meristems. *Plant J.*, 5, 571–582.

Mathur, A.K., Ahuja, P.S., Pandey, B. and Kukreja, A.K. (1989) Potential of somaclonal variations in the genetic improvment of aromatic grasses. In A.K. Kukreja, P.S. Ahuja and R.S. Thakur (eds.), *Tissue Culture and Biotechnology of Medicinal Plants*, CIMAP, Lucknow, pp. 78–89.

Mucciarelli, M., Gallino, M., Scannerini, S. and Maffei, M. (1993) Callus induction and plant regeneration in *Vetiveria zizanioides*. *Plant Cell Tiss. & Org. Cult.*, 35, 267–271.

Mucciarelli, M., Cozzo, M., Gallino, M., Scannerini, S., Bocco, A., Brazzaventre, S. and Maffei, M. (1994) XAD-4 elicitation effects on cell growth and sugar uptake in cell suspension cultures of *Vetiveria zizanioides* Stapf. (Poaceae). *Giornale Botanico Italiano*, 128, 1, 294.

Mucciarelli, M., Maffei, M., Gallino, M., Cozzo, M. and Scannerini, S. (1997) *Vetiveria zizanioides*: as a Tool for Environmental Engineering. *Acta Horticulturae*, 457, 261–270.

Müller, A.J. and Grafe, R. (1878) Isolation and characterization of cell lines of *Nicotiana tabacum* lacking nitrate reductase. *Physiol. Plant.*, 15, 473–497.

Murashige, T. and Skoog, F. (1962) A revised medium for rapid growth and bioassays with tobacco tissue cultures. *Physiol. Plant.*, 15, 473–497.

Ozias-Akins, P. and Vasil, K. (1982) Plant regeneration from cultured immature embryos and inflorescences of *Triticum aestivum* L. (wheat): evidence for somatic embryogenesis. *Protoplasma*, 110, 95–105.

Patnaik, J., Sahoo, S. and Debata, B.K. (1997) Somatic embryogenesis and plant regeneration from cell suspension cultures of palmarosa grass (*Cymbopogon martinii*). *Plant Cell Reports*, **16**, 430–434.

Peyron, L. (1995) Le vetyver et sa culture dans le monde. *Rivista Italiana EPPPOS*, **17**, 3–18.

Rao, A.M., Kavi Kishor, P.B. and Anada Reddy, L. (1988) Callus induction and high frequency plant regeneration in Italian millet (*Setaria italica*). *Plant Cell Rep.*, 7, 557–559.

Reisch, B. (1983) Genetic variability in regenerated plants. In D.A. Evans, W.R. Sharp, P.V. Ammirato and Y. Yamada (eds.), *Handbook of Plant Cell Culture*, Macmillian Publishing Company, New York, pp. 748–769.

Schwendiman, J., Pannetier, C. and Michaux-Ferriere, N. (1988) Histology of somatic embryogenesis from leaf explants of the oil palm *Elaeis guineensis*. *Ann. Bot.*, **62**, 43–52.

Sethi, K.L. and Gupta, R. (1980) Breeding for high essential oil content in Khus (*Vetiveria zizanioides*) roots. *Indian Perfumer*, **24**, 72–78.

Shasany, A.K., Lal, R.K., Khanuja, S.P.S., Darokar, M.P. and Kumar, S. (1998) Comparative analysis of four genotypes of *Vetiveria zizanioides* through RAPD profiling. *J. Med. Arom. Plant Sc.*, **20**(4), 1022–1025.

Skoog, F. and Miller, C.O. (1957) Chemical regulation of growth and organ formation in plant tissue cultured *in vitro*. *Symp. Soc. Exp. Biol.*, **11**: 118.

Sreenath, H.L. (1983) Cytogenetic and tissue culture studies in some important species of aromatic grasses. PhD. Thesis, Mysore University, Mysore.

Sreenath, H.L. and Jagadishchandra, K.S. (1989) Somatic embryogenesis and plant regeneration from the leaf cultures of Vetiveria zizanioides (Khus-khus grass). *XIII Plant. Tissue Culture Conf. Dep. Bot. North-Eastern Hill Univ.*, Shillong, p. 54, 1989.

Sreenath, H.L. and Jagadishchandra, K.S. (1990) Propagation of *Vetiveria zizanioides* (L.) Nash (Khus grass) through inflorescence and node culture. *Palai J.*, **12**(2), 27–29.

Sreenath, H.L., Jagadishchandra, K.S. and Bajaj, Y.P.S. (1994) *Vetiveria zizanioides* (L.) Nash (Vetiver Grass): In vitro culture, regeneration, and the production of essential oils. In Y.P.S. Bajaj (ed.), Biotechnology in Agriculture and Forestry 26, *Medicinal and Aromatic Plants VI*, Springer-Verlag, Berlin, pp. 403–421.

Stepan-Starkissian, G. and Fowler, M.W. (1986) The metabolism and utilization of carbohydrates by suspension cultures of plant cells. In M.J. Morgan (ed.), *Carbohydrate Metabolism in Cultured Cells*, Plenum, New York, pp. 151–181.

Strobel, J., Hieke, M. and Gröger, D. (1991) Increased anthraquinone production in *Gallium vernum* cell cultures induced by polymeric adsorbents. *Plant Cell Tiss. & Org. Cult.*, **24**, 207–210.

Timmers, A.C.J., De Vries, S.C. and Schel, J.H.N. (1989) Distribution of membrane-bound calcium and activated calmodulin during somatic embryogenesis of *Daucus carota* L. *Protoplasma*, **152**, 24–29.

Tisserat, B. (1985) Embyogenesis, organogenesis and plant regeneration. In R.A. Dixon (ed.), *Plant Cell Culture. A Practical Approach.*, IRL Press, Oxford, pp. 79–105.

Vasil, I.K. and Vasil, V. (1986) Regeneration in cereal and other grass species. In I.K. Vasil (ed.), *Cell Culture and Somatic Cell Genetics of Plants*, Academic Press Inc., Orlando, pp. 121–150.

Vasil, I.K. and Vasil, V. (1991) Embryogenic callus, cell suspension and protoplast cultures of cereals. In I.K. Vasil (ed.), *Plant Tissue Culture Manual*, Kluwer Academic Publishers, Dordrecht, Netherlands.

Wernicke, W. and Brettell, R. (1980) Somatic embriogenesis from *Sorghum bicolor* leaves. *Nature*, **287**, 138–139.

Whiters, L.A. (1985) Cryopreservation and Storage of Germplasm. In R.A. Dixon (ed.), *Plant Cell Culture. A Pratical Approach*, IRL Press, Oxford, pp. 169–191.

8 Economic Importance, Market Trends and Industrial Needs, and Environmental Importance

Michael W.L. Pease

Co-ordinator, European and Mediterranean Vetiver Network. Quinta das Espargosas, Odiaxere, 8600 Lagos, Algarve, Portugal.

Introduction

For several centuries vetiver grass has been cultivated commercially for the aromatic and medicinally valued oil that can be distilled from its roots. In the late 19th and early 20th centuries manufacturing industries were established in such locations as Haiti, Reunion (Seychelles) and Java (Indonesia) and subsequently in Louisiana (USA) and Brazil (TVN Newsletter, 1998). These industrial producers are relatively small-scale manufacturers since the global market for the essential oil that is produced may be little more than 250 tons per annum. It was not until the 1980s that *Vetiveria zizanioides* (vetiver grass) was fully recognised for its value as a vegetative barrier able to control soil erosion, contain water run-off and reduce loss of plant nutrients. It is in these areas together with its value as a tool in bio-engineering that its primary economic benefits now lie. A technology known as Vetiver Grass Technology (VGT) has been developed for application in these roles but this is a relatively recent phenomenon. Consequently little detailed information is available regarding its economic benefits most of which do not have a cash value. Some location-specific data is available on the costs of nursery establishment, field planting and annual maintenance but, even in this regard, much still needs to be determined. The value of economic benefits, which are at best difficult to quantify, are generally unknown. What follows is, therefore, a general discussion of the economic background to VGT, rather than a detailed economic analysis.

Costs

Vetiver grass is a low cost, economic system of soil and moisture conservation. Economic analyses, conducted in the early 1990s compared the establishment of vetiver grass hedges at less that US$30/ha with the more than US$500/ha for conventional engineered terraces (Grimshaw and Helfer, 1995). Economic rates of return for the latter were around 20% compared to more than 90% for vetiver.

In Sichuan, China, soil conservation measures involving bunded terraces cost five times more than vetiver grass hedges (Zhang, 1992). A similar ratio has been measured in India where the cost of bunds was US$60 per hectare and for vetiver less than US$20 per hectare (National Research Council, 1993).

Also in China it was shown that the cost of highway and railroad cut and fill slopes with vetiver is 10% of the cost of stone-based technologies (HuiXiu *et al.*, 1997).

Table 8.1 Economic benefit of using vetiver, cost-benefit ratios.

Treatments	Sorghum	Cotton	Mung Bean	Sorghum Mung Bean	Cotton Mung Bean	Pooled Mean
Across the slope	1.84	2.26	0.68	2.29	2.69	1.95
Vetiver hedgerows	2.48	2.76	1.22	3.41	3.79	2.73
Graded bunds	1.26	1.54	0.44	1.60	1.98	1.36
Mean	1.86	2.19	0.78	2.43	2.82	

Table 8.2 Daily Cost of Labour per hectare (in US$ equivalent, see text for explanation).

Slope	$1.00	$2.00	$3.00
2–5%	24.08	38.21	52.34
5–10%	51.60	81.88	112.17
20–30%	172.01	272.95	373.89
50–60%	378.41	600.48	822.55
90–100%	653.62	1,037.19	1,420.77

Table 8.3 Comparison between the cost of producing vetiver slips by direct planting in plastic bags and by replanting in plastic bags following the first step.

	In Plastic Bags (US$)	In The Field (US$)	In Greenhouses (US$)
First Step – Direct planting	0.0048	0.0032	0.0080
Replanting in plastic bags subsequent to first step	0.0660	0.0644	0.0692

Table 8.1 shows benefit:cost ratios and demonstrates the economic benefit of using vetiver hedgerows planted on the contour as against across the slope planting or graded bunds (Bharad, 1993; Bharad and Bathkal, 1993; Sagare and Meshram, 1993). Over time "across the slope" production results in severe soil erosion and crop reduction; graded bunds wear down over time and require expensive reconstruction.

The cost of land treatment with contour hedges of vetiver grass increases with slope gradient and labour cost per man-day (World Bank Handbook, 1988). Planting of steeper slopes requires greater hedgerow frequency and, because the work is more difficult, results in increased number of man-days per linear metre. Table 8.2 gives some examples as to how these parameters change.

In 1996 in Thailand, comparison was made between the cost of producing vetiver slips firstly by direct planting in plastic bags and secondly by replanting in plastic bags following the first step (Table 8.3) (Sumol et al., 1996).

In 1988 in India, a per hectare budget was produced for the development of a vetiver grass nursery based on the actual costs of a government operated nursery (Grimshaw and Helfer, 1995). Total rounded costs amounted to approximately US$820 equivalent. With an output of 60 slips per clump a farmer could expect a gross margin of approximately US$1,680 equivalent. A similar calculation was made in China in 1989 giving a total rounded cost of between US$1,020 and US$1,500 equivalent per hectare, dependent upon regional labour costs (Grimshaw and Helfer,

Table 8.4 Gross margins calculated for nurseries with two alternative production outlets.

1 ha Nurseries	Gross Margin US$s equivalent
Model 'A' – Selling bare root and containerised stock	12,256
Model 'B' – Selling bare root stock only	2,126

1995). Output was estimated at approximately US$4,800 giving a gross margin of between US$3,300 and US$3,800 equivalent.

From experience in Thailand and Honduras, for example, a farmer can plant about 100 linear metres of hedge per day under difficult conditions or 200 metres under easy conditions (Leonard, 1995). This compares with 5–10 metres per day of rock wall or contour ditch construction.

Field costs of establishing vetiver hedges were calculated in India in 1988 at US$18.40 equivalent per hectare (Grimshaw and Helfer, 1995).

More recent nursery development costs for Madagascar give a total cost of US$1,430 equivalent per hectare to produce some 1,111,000 slips per annum. The cost per marketable slip of 3 tillers was estimated at US$0.002 (rounded up). Gross margins were calculated for nurseries with two alternative production outlets (Table 8.4) (Juliard and Grimshaw, 1998).

Field costs for establishing hedges using bare-rooted vetiver slips per 100 metres were calculated at US$11.55 equivalent per linear metre.

Benefits

Benefits of VGT can be divided between those having cash value, immediate substitution value, immediate cost savings or unquantified long-term environmental impact. It is in the latter classification that the greatest economic value exists. Because these are, as yet, unquantified the economic value of VGT remains inadequately recognised.

Cash value

Essential oil

Oil is distilled from the roots of vetiver grass with steam, quality being dependent upon the age of the root material and the method of distillation. The extraction process may take as long as 72 hours but is generally about 24–36 hours, about 15–20 of which are for actual distillation. Distillation is somewhat difficult since the oil does not readily separate from the water (see Chapter 3). The chief volatile constituents have a high viscosity with high boiling points (see Chapter 4). One hectare of vetiver plants will yield about 1,000 kg of air-dried root per hectare under good soil conditions. The yield of extracted oil is about 1.5 to 2.0% and reportedly up to 2.5% in China (Hanping, 1997). Oil production is therefore about 15 to 20 kg per hectare. Young, thin roots yield light oil with low specific gravity and with poor solubility. Such oils have a green, earthy and rather harsh aroma. Older roots yield better quality oil with a higher specific gravity and optical rotation levels. These oils are darker in colour (reddish-brown) and the aroma is fuller, richer and more long lasting. Quality is determined also by the length of the distillation process. Since the most valuable constituents of the oil and the most important in regard to aroma have high boiling points they can only be recovered by prolonged distillation, resulting in higher processing costs. Costs can be reduced somewhat by using superheated steam towards the end of the distillation process. The optimum time for harvesting occurs some 24 months following planting. The roots can be distinguished by having fine black rings as against the dead, grey-brown colour of older roots. Early harvesting results in somewhat increased quantities of extractable oil but of lower quality. Late harvesting results in sub-economic extractable oil yields. The aroma of vetiver oil improves with age. Oil from Haiti is of good quality and aroma and has a different classification to that from Java or Reunion, for example. Global demand for vetiver oil is limited to about 250 tons per annum (Grimshaw and Helfer, 1995). The principal importers of oil, for the production of scents are the United States, France, Switzerland and the United Kingdom. (For more information on essential oil production see Chapters 1, 3 and 4.)

Handicrafts

Vetiver grass is valued by rural communities for its long-lasting usefulness as a plaiting grass in the production of such items as mats, hats, rope and mullen, baskets, belts, combs, fans, hairbrushes, lamp shades, sandals, toothpicks and fire-lighters. The value of these by-products is difficult to quantify and, in any event, is small. However, they can represent important small-scale industries in rural communities having both a cash and social value and aiding in the preservation of the social fabric (Grimshaw and Helfer, 1995; TVN Newsletter, 1996). For handicraft pictures see Chapter 3.

Domestic and medicinal use

Rural communities have discovered over time a wide-ranging domestic and medicinal application for both the leaves and roots of vetiver grass. Domestically these include leaf infusion as a tea, curry seasoning, meat spice, root aroma in drinking water, sachets filled with ground root as a pleasant aroma and to deter moths from hanging clothes and as an insect repellent, notably against fleas. Medicinal extracts, generally from the roots, are prepared in many countries and purportedly control a variety of diseases, fevers and medical ailments (see also Chapter 5 for some pharmacological properties). Vetiver is also used as a sleep inducer, to calm nerves and induce sweating in people and animals, and as an enema. The plant is therefore prized by rural communities and aids rural community stability (TVN Newsletter, 1995, 1996).

Pulp production for paper manufacturing

The stem and leaves of vetiver are reported to be suitable for pulp and paper production and for writing and printing papers, but economic data was not available (Grimshaw and Helfer, 1995).

Crop yield enhancement

Planting of vetiver hedges can result in some decreased yield in the row of a crop growing immediately adjacent to the hedge. This can be reduced by trimming the hedge to allow strong early growth of the inter-crop (Kon and Lim, 1991; Grimshaw and Helfer, 1995). The establishment of vetiver hedgerows results in the build up of natural terraces of top soil between hedges, increased availability of precipitation for plant uptake by crops grown between the hedgerows and reduction in loss through run-off of both organic and inorganic plant nutrients. These beneficial factors far outweigh any loss in yield that may occur in that part of a crop growing immediately adjacent to the vetiver hedgerow. This is generally confined only to the first immediately adjacent row although, in Australia, under dryland conditions negative effects of the vetiver hedge were noted on the first two rows in the case of sorghum (Dalton and Truong, 1996). The Centro Internacional de Agricultura Tropical (CIAT) conducted trials in 1990–1992 in Colombia on cassava grown under different soil conservation protective regimes including vetiver. These trials indicated a net increase in production of fresh cassava roots of 0.8 tons/ha where vetiver hedges were used (Lang and Ruppenthal, 1992). In the case of sugar cane in Fiji the provision of vetiver hedges resulted in improved ratooning (7–8 crops) and cane yields were improved by 48 t/ha or 55% (TVN Newsletter, 1995). Reports from Java, Indonesia indicated that maize and bean yields increased by 104.7% and 142.9% respectively (Donie and Sudradjat, 1996). Overall, crop yields can increase by as much as 40% at a cost that may be no more than about US$3 per hectare.

Fruit Tree Yield Enhancement

Fruit tree yields can be enhanced up to 20% by planting vetiver hedges in semi-circles on the downside of the trees to conserve soil moisture.

Immediate substitution value (quantifiable)

Livestock fodder

Vetiver leaves are palatable and attractive to most herbivores, particularly in an early stage of growth. In Lesotho it was found that 0.40 ha vetiver was sufficient nutrition for 1 cow and 1 heifer for up to 60 days (Chaudhry and Petlane, 1995). The nutritive value of freshly cut vetiver grass in its green leaf stage is between Napier grass (*Pennisetum purpureum*) and fresh corn stover (National Research Council, 1993; Chaudhry and Petlane, 1995). The International Livestock Center for Africa (ILCA) records vetiver as yielding 2–10 tons DM/ha with a digestibility coefficient of 35–40% (National Research Council, 1993). In Thailand, it was shown that young vetiver grass leaves were suitable for ensiling providing a good quality, palatable and digestible silage with good percentages of pH and DM (Chaudhry and Petlane, 1995).

Mulch and substrate

The trimmings from vetiver hedges provide a good quality, long-lasting, mulch that breaks down slowly and has high absorptive value (Wondimu, 1990). Analysis of mulch shows that it contains considerable quantities of nitrogen, phosphorus, potassium and magnesium (Yoon, 1991; National Research Council, 1993). Vetiver mulch is of special value to tree crops and reduces rain-drop impact on soil surfaces when used as a further measure of soil conservation. Vetiver mulch has been shown to be particularly suitable as substrate for mushroom production in Thailand and China (Saifa *et al.*, 1996).

Fish food

Vetiver planted around fish-ponds reportedly provides a useful source of food for carp (Welsin, 1991; National Research Council, 1993).

Fuel briquettes

In the USA and UK, in particular, work has been conducted into the making of fuel briquettes from vetiver leaves. No data was identified relative either to calorific or economic values (TVN Newsletter, 1995).

Livestock bedding

Because of their good absorptive value and long-lasting strength, vetiver leaves make good bedding for livestock (Grimshaw and Helfer, 1995).

Shade value for lambs

A vetiver hedge provides protection for young lambs from heat stress and wind blow. In Queensland, Australia, it was shown that overall productivity of sheep increased by 15% when protection from these elements was provided by vetiver hedges (Somes and Bortolussi, 1994; TVN Newsletter, 1995).

Thatch for roofing material

From experience in African countries, in particular, vetiver grass is a valuable source of thatching material resisting rot and lasting longer than many other grasses (TVN Newsletter, 1995).

Mattress stuffing

In many African countries, in particular, vetiver grass is valued for mattress stuffing because of its insect repelling and long-lasting characteristics (National Research Council, 1993).

Insect repellent

The odour from vetiver leaves and from ground up roots is repellent to many insects including fleas, flies, cockroaches, clothes moths, lice and bedbugs (National Research Council, 1993). Thus vetiver leaves used as thatch or mattress stuffing, or roots when ground into powder and held in sachets are valued domestically.

Biomass production

Vetiver has been grown for biomass production although, in this respect, higher yielding alternatives such as Napier Grass are available. In the USA it was found that biomass production from 6 month old vetiver grass ranged from 123,516 kg/ha to 358,439 kg/ha for fertilised and unfertilised plots. The calorific value of vetiver is about 2,700 btu/kg. In Texas under irrigated conditions production of more than 100 tons DM has been achieved, equivalent to about 350 tons fresh leaf (Igbokwe *et al.*, 1991; Grimshaw and Helfer, 1995).

Immediate cost savings

Stabilisation of engineered structures i.e. roads, highways, culverts, bridges and dams

Alternative vegetative means of stabilising sensitive slopes and engineered structures using VGT as against engineered means has gained increasing recognition in recent years. VGT can be seen as one of a number of tools that can be used independently or in conjunction with others to reduce costs and effectively stabilise or protect expensive engineered structures. For instance, in China the cost of an engineered rock wall to protect the steep slope of a highway cutting cost US$3 equivalent per square metre, compared to US$0.3 per square metre for vetiver hedges, i.e. one tenth the cost. Corroborative figures in Australia, with higher labour costs, were US$100 equivalent for a rock wall compared to US$10 equivalent for vetiver hedges (TVN Newsletter, 1997). Such engineered features as bridges, abutments and culverts are especially sensitive to undercutting and cave in from erosion of surrounding soils. Vetiver hedges provide inexpensive protection. Engineered methods of land stabilisation can include such methods as blown gunite, gabions, concrete drainage ways, and man-made fibre substances. All of these can either be substituted by or used in conjunction with VGT at significant cost saving advantage.

Stabilisation of dam walls

Earthen-built dams frequently use highly erodable soils in their construction. If left unprotected, dam walls are likely to need major mechanical repair within a few years of construction. The use of vetiver hedges to stabilise and protect dam walls can lead to direct savings by avoiding the need for expensive, mechanical repair.

Unquantified long-term environmental benefits

Control of soil erosion

The extent of soil loss globally is immense. The Food and Agriculture Organization (FAO) estimated this at some 20 billion tons per annum, equivalent to some 5–7 million hectares of good land (National Research Council, 1993). Its economic value to global food security and overall environmental protection is difficult to measure. The use of a vetiver grass hedge can reduce soil erosion by 50–70%. In physical terms, a vetiver hedge will reduce soil loss from about 8–11 tons per ha. to 2–3 tons per ha. (Grimshaw and Helfer, 1995; Donie and Sudradjat, 1996).

Control of water run-off and groundwater recharge

It has been estimated that water run-off can be reduced by 50–70% where VGT is employed (Donie and Sudradjat, 1996; Grimshaw and Helfer, 1995). The water that is retained firstly benefits plant growth leading to direct increase in yields. Secondly, in the longer term, deep penetration results in replenishment of subterranean water resources. There is, thus, an important and significant long-term environmental impact arising from the use of these vegetative barriers. Measurements in the United States show that decreases in run-off from grass hedged fields compared to fields without hedges range from 10–70% (Kemper, 1999). Furthermore, in the case of vetiver hedges their height reduces wind velocity thus having a beneficial effect on evapotranspiration. Because of the nature of their deep rooting system vetiver plants tend to draw their needs from deeper sources than in the case of most other plants.

Reduction in loss of plant nutrients

FAO has calculated that, on average, most farmers lose to run-off some 50% of the fertilisers applied (National Research Council, 1993). In the United States, for example, it is estimated that some 18 billion tons of fertiliser is lost per annum in run-off (National Research Council, 1993). Loss of organic fertiliser is equally alarming. The use of VGT impedes the flow of water run-off, traps plant nutrients behind the hedges and holds them for plant use in the naturally formed terraces that result.

Reduction in silting of dams and rivers

As an example of the extent of silting of dams in Morocco it is estimated that one 150 cu. metre dam is required annually simply to compensate for those lost to siltation (National Research Council, 1993). Vetiver grass barriers can significantly reduce the direct silt flow that enters dams and if VGT is used extensively in the surrounding watershed, waters flowing into dams will contain significantly reduced quantities of silt. The long-term economic benefits of using vetiver to reduce siltation of dams are therefore large.

Control of leachate and heavy metal toxins from industrial and municipal waste dumps

Vetiver grass has been demonstrated to be tolerant to higher levels of heavy metal toxins than most other plants (Hengchaovanich, 1998; Truong and Baker, 1998). Furthermore, in the case of most heavy metals they are stored in the roots with only small proportions passed to the leaves. Typically, such heavy metals occur in waste dumps created by industry or by municipalities. If inadequately protected, leachate seeps from these dumps to pollute nearby water courses. Vetiver planted around such dumps filters the leachate and provides considerable protection against water pollution. Economic impact can be measured in terms of the inexpensive improvement to water supplies and the reduction in expensive chemical and mechanical filtration that would otherwise have to be undertaken.

River bank stabilisation and drainage way protection

A frequent location for erosion is along the banks of water-courses. Typically, such water-courses pass through cultivated valley bottoms, valuable for agriculture. Vetiver grass thrives on moist conditions and periodic flooding and the deep roots stabilise banks and protect against encroachment into cultivated areas alongside. Similar protection can be provided by vetiver grass for drainage ways which, if sufficiently narrow, can be shielded from weed growth and evaporation by vetiver hedges planted on either side (TVN Newsletter, 1997). Economic benefits result from limitation of loss of productive land and cleaner water supplies.

Boundary demarcation

Boundary demarcation with stone walls, fencing or other similar means is costly, particularly in labour, and often has a relatively short life-span without regular and expensive maintenance (Grimshaw and Helfer, 1995). Vetiver hedges used for boundary demarcation are permanent, relatively low-cost for both establishment and maintenance, cost effective and environmentally friendly. A well-planted dense hedge will also reduce ingress by animals. Vetiver grass boundary hedges can therefore be regarded as a cost-saving tool in this regard.

Stabilisation of dam spillways, surrounds and inside of walls against wind and lap erosion

Dams are expensive to construct and unless adequately maintained will deteriorate over time to the point where they become ineffective or dangerous. Deep gullies can develop up spillways, surrounds deteriorate and the insides of walls become worn down with wind and lap erosion. Vetiver grass barriers can provide protection from these factors resulting in economic benefits in increased dam life.

Stabilisation of urban or industrial development lands

In the course of urban and industrial construction existing soils are disturbed and become liable to erosion. VGT protects such engineered areas and provides an aesthetically pleasant and controlled environment. Benefits accrue from reduced maintenance costs on engineered structures and from the improved environmental atmosphere that is created.

Stabilisation of dune lands and cliff tops from landward-side erosion

The direct benefits of protecting dune lands from seaward-side erosion and cliff tops from landward-side erosion lie largely within the industry of tourism. However, the overall environmental impact of undertaking these conservation measures, though difficult to quantify, is nevertheless of considerable importance. Vetiver grass is tolerant to relatively high levels of salinity (Truong et al., 1991; Grimshaw and Helfer, 1995).

Eutrophication of polluted waters, notably from inorganic fertilisers,
pesticides, animal manures and algae build-up

An example of serious, large-scale pollution of waters occurs in Lake Taihu, China which covers some 2,420 sq. km. Water eutrophication is evident in high levels of nitrogen, phosphorus and carbon elements in both the water and nearby soils, resulting in rapid growth of algae. The nitrogen is mostly concentrated along the lake edges where it is particularly suitable for planting vetiver. Since the vetiver has a high uptake of nitrogen it acts as a water purifier. An additional technology is to establish the vetiver on floating bamboo platforms which can remove 99% of water soluble phosphates and 82% of total nitrates (Hanping *et al.*, 1997; Zhen *et al.*, 1997). The benefits that result from such actions are in improved human and animal use of such waters and in the build up of fish stocks.

As a pioneer plant on disturbed lands bare of vegetation, encouraging ingress
of indigenous species

Vetiver has been established in such locations as the slimes surrounding mining sites in South Africa and coal mine tailings in Australia which have hitherto been devoid of plant growth (Berry, 1996; Tantum, 1996). Once established, the vetiver acts as a pioneer plant creating a micro climate under which indigenous species can and do thrive. The economic value of vetiver in such circumstances lies in its ability to rehabilitate lands that would otherwise be useless and environmentally unattractive.

Conclusions

Vetiver grass is a versatile tool in addressing a number of serious environmental problems. It possesses a number of attributes that can generate income including essential oil commerce but, more significantly, it can contribute significantly towards long-term environmental benefits by reducing soil erosion, retaining soil moisture for increased plant growth, increasing subterranean water supplies and reducing loss of plant nutrients. Vetiver Grass Technology can significantly reduce costs in such bio engineering contexts as road and highway slope protection and dam protection. VGT has important economic and social benefits to rural communities where it can be a source of income and can also underpin the social fabric through the protection of the conditions of production under which such communities live. Overall, VGT can be regarded as an important tool for protecting the means of survival of the world's increasing populations.

References

Berry, M.P.S. (1996) The Use of Vetiver Grass in Revegetation of Kimberlite soils in Respect to South African Diamond Mines. *TVN Newsletters*, 15.

Bharad G.M. and Bathkal, B.C. (1991) Extracts from Role of Vetiver Grass in Soil and Moisture Conservation. *TVN Newsletters*, 6.

Bharad, G.M. (1993) Experiences with Vetiver in Maharashtra, India. *TVN Newsletters*, 10.

Chaudhry, A.B. and Petlane, M. (1995) Prospects And Problems Associated With The Use Of Vetiver Grass As A Biological Mean Of Soil And Water Conservation In Lesotho Lowlands And Foothills. *TVN Newsletters*, 14.

Dalton, P.A. and Truong, P.N.V. (1996) Soil Moisture and Sorghum Yield as Affected Vetiver Hedges under Irrigated and Dryland Conditions. *TVN Newsletters*, 15.

Donie, S. and Sudradjat, R. (1996) Vetiver Grass as Erosion and Land Productivity Control. *TVN Newsletters*, 15.

Grimshaw, R.G. and Helfer, L. (1995) *Vetiver Grass for Soil and Water Conservation, Land Rehabilitation, and Embankment Stablization: A Collection of Papers and Newsletters Compiled by the Vetiver Network*. World Bank technical paper, ISN 0253–7494; no 273.

Hanping, X. (1997) Observations and Experiments on the Multiplication, Cultivation, and Management of Vetiver Grass Conducted in China in the 1950's. *TVN Newsletters*, 18.

Hanping, X., Huixiu, A., Shizhong, L. and Daoquan, H. (1997) A Preliminary Study on Vetiver's Purification for Garbage Leachate. *TVN Newsletters*, 18.

Hengchaovanich, D. (1998) *Vetiver Grass Slope Stabilization and Erosion Control*, Tech. Bull.No 1998/2 Pacific Rim Vetiver Network (PRVN), Office of The Royal Projects Development Project Board (RDPB), Bangkok, Thailand.

HuiXiu, A., HanPing, X., ShiZhong, L. and DaoQuan, He (1997) Studies on protecting highway slopes with vetiver hedgerows, *TVN Newsletters*, 18.

Igbokwe, P.E., Tiwari, S.C., Burton, J.E. and Waters, R.E. (1991) Influence of Accession Variability and Fertilisation on the Establishment and Growth of Vetiver Grass in a Field Nursery. *TVN Newsletters*, 7.

Juliard, C.J. and Grimshaw, R.G.G. (1998) Personal Communication.

Kemper, D. (1999) Personal Communication.

Kon, K.F. and Lim, F.W. (1991) Vetiver Research in Malaysia: Some Preliminary Results on Soil Loss, Runoff and Yield. *TVN Newsletters*, 5.

Lang, D. and Ruppenthal, M. (1991) Reports on the Second Year's Results in Cassava Systems with Living Barriers of Vetiver Grass and Elephant Grass. *TVN Newsletters*, 8.

Leonard, D. (1995) The Land Use Productivity and Enhancement Project's Experiences with Vetiver Barriers in Honduras. *TVN Newsletters*, 13.

National Research Council (1993) *Vetiver Grass: A Thin Green Line Against Erosion*. National Academy Press, Washington, D.C.

Sagare, B.N. and Meshram, S.S. (1993) Evaluation of Vetiver hedgerows relative to graded bunds and other Vegetative hedgerows. *TVN Newsletters*, 10.

Saifa, Y., Taptimorn, P. and Pitkpaivan, P. (1996) Vetiver Grass (*Vetiveria nemoralis*) As Substrate for Mushroom Cultivation. *TVN Newsletters*, 15.

Somes, T. and Bortolussi, G. (1995) Asian Grass to Lift Lambing Percentages. *TVN Newsletters*, 13.

Sumol, S., Wichai, S. and Darunee, K. (1996) Cost Comparison in Producing Vetiver Grass from Different Methods of Production. *TVN Newsletters*, 15.

Tantum, A. (1996) Vetiver in a Southern African Context. *TVN Newsletters*, 15.

Truong, P.N., Gordon, I.J. and McDowell, M.G. (1991) Excerpts from Effects of Soil Salinity on the Establishment and Growth of Vetiveria zizanioides (L) Nash. *TVN Newsletters*, 6.

Truong, P.N.V. and Baker, D.E. (1998) *Vetiver Grass System for Environmental Protection*. Tech. Bull No 1998/1, Pacific Rim Vetiver Network (PRVN) Office of the Royal Projects Development Board (RDPB), Bangkok, Thailand.

TVN Newsletter – The Vetiver Network Newsletters published biannually by The Vetiver Network, 15 Wirt St. N.W., Leesburg, VA 20176, USA (ftp://www.vetiver.org/Newsletters/Nlcontents.htm).

Welsin, G. (1991) Contribution to *TVN Newsletters*, 7, in Grimshaw, R.G. and Helfer, L. (1995).

Wondimu, M. (1990) Contribution to *TVN Newsletters*, 4, in Grimshaw, R.G.G. and Helfer, L. (1995).

World Bank Handbook (1988) *Vetiver Grass – A Method of Vegetative Soil and Moisture Conservation*. The World Bank, Washington, D.C.

Yoon, P.K. (1991) *A look-See at Vetiver Grass in Malaysia: First Progress Report.* Unpublished Report.

Zhang, X. (1992) Vetiver Grass in Peoples Republic of China. *TVN Newsletters,* **8**.

Zhen, C., Tu, C. and Chen, H. (1997) *Preliminary Experiment on Purification of Eutrophic Waters with Vetiver. TVN Newsletters,* **18**.

9 Beyond the Vetiver Hedge

Organizing Vetiver's Next Steps to Global Acceptance

Noel Vietmeyer

Office of International Affairs. National Academy of Sciences.
Washington, DC 20418, USA

Introduction

Despite its great merits vetiver has a major problem: it just is not going to make many millionaires. Were it capable of creating lots of money for individuals, vetiver specialists could retire to their laboratories, offices and test plots strong in the conviction that others would eagerly turn all their results into practical benefits.

But the reality is that people of extraordinary conviction and vision, such as His Majesty the King of Thailand, are the only ones going to dedicate their energies to moving vetiver upward and outward to its global destiny. As a result, one cannot go back to research and expect that this immensely useful plant will advance into widespread acceptance by some sort of global osmosis. Vetiver champions must now shoulder the burden of selling the vetiver idea to people of influence worldwide.

This brings up a second problem: vetiver is so good at doing so many things that the immediate challenge is an organizational one. Even vetiver specialists become overwhelmed by the sheer breadth of the possibilities envisaged. And if specialists are confused, consider how baffling the story must be to those newcomers who must be brought on board to achieve a successful global outcome.

To help bring some measure of order to the collective vision, as well as to boost the crop's advancement, the following initiatives are essential to move the case for vetiver forward: –

- Soil-Erosion Initiative;
- Extreme-Soil Initiative;
- Water-Management Initiative;
- Pollution-Control Initiative;
- Farmer-Support Initiative;
- Disaster-Prevention Initiative;
- Basic-Science Initiative.

These initiatives are more than just ways of thinking about the plant and its promise, they are compartments of practical progress, each distinct and self-contained within itself. Of course there are overlapping borders – indeed, a complicated chart could be drawn showing all the interrelationships but for all that, each of these results-oriented topics plays on different strengths of the grass and reaches out to different audiences.

In addition, each requires different actions as the burden of locating partners for mutual support and for faster progress worldwide is shouldered.

Initiatives

The Soil-Erosion Initiative

Of all vetiver's applications, controlling soil erosion is by far the best understood and furthest advanced. Probably 90 percent of all the work to date has been devoted to this initiative, and the fact that the plant stops soil loss is now abundantly clear. The effect is due largely to the strength of the stems in hedges placed along the contours of hillslopes.

The progress of this initiative must not be slackened. Soil erosion is arguably the worst global environmental problem, and for much of the world it was the least tractable until vetiver came along. All in all, this grass offers the first practical intervention with worldwide possibilities.

The Soil-Erosion Initiative's next major challenge is to project existing knowledge to new locations and new people. In a sense, it is essential to bring other nations up to the level of commitment and action achieved here in Thailand. If perhaps 100 more nations can be made as committed as this one, the global scourge of soil erosion would mostly be thwarted within our lifetimes. Of course, some nations are too frozen during winter to consider vetiver, but the United States is developing a complementary, vetiver-inspired, grass-hedge technology using cold-climate species.

Bringing about the tantalizing vision of global success against erosion should be the Soil-Erosion Initiative's aim. The existing vetiver publications are, by and large, adequate to the task. Farmers and foresters are of course the main audience, but it is necessary to reach out more to engineers and persuade them to take up vetiver routinely along roadsides, around construction sites, next to bridge abutments and along pipelines. Also it is essential to reach city officials so that vetiver is put to use stopping erosion in the squatter settlements, stormwater drains and other urban sites.

In addition, environmental scientists and conservation watchdog groups need to be made aware that vetiver is now a promising answer to the soil that washes into natural preserves. They could, for example, push for the regional employment of vetiver hedges to reduce the water-borne silt that devastates coral reefs, fish-spawning grounds and various other irreplaceable habitats. Three examples worth vigorous action are listed below.

The everglades

The delicate balance of this irreplaceable habitat in Florida is being upset by phosphate and other nutrients washing out of nearby sugarcane fields. To me, the solution lies in surrounding the canefields with vetiver hedges. Those hedges would trap the silt (along with the phosphate clinging to it) and absorb soluble nutrients before the water ever passes into the Everglades.

Lake Victoria

This large lake in the heart of Africa is suffering explosive blooms of water hyacinth. I am informed that the problem has been linked to nutrient-laden silt washing off the land and fertilizing the weed. A regional vetiver-planting campaign in the watersheds serving the problem locations might immeasurably benefit the lake, not to mention the watersheds themselves.

The East African coast

The grass might also prove useful in watersheds in eastern Kenya, where silt washing off the land is killing a priceless coral reef.

The Extreme-Soil Initiative

The primary challenge in this initiative is not erosion control; it is to make extreme soils productive, or at least more productive than at present. This is also an important challenge. Vast areas of the earth typically classified as "marginal lands," "waste lands," or "abandoned lands" are inadequately used because they are just too hard to harness for crop production.

A truly amazing aspect of vetiver is its ability to survive on sites so hostile toward plant life that people now universally write them off as impossible to cultivate. The relevant feature in this case has to do with the plant's root chemistry. It is known from experiments and observations that vetiver grows in acid soil, alkaline soil, laterite, vertisol, toxic mine spoil, moderately saline soil, wetland and dryland soil, and even soils so dense they are likened, not inaccurately, to "concrete."

That vetiver can survive in such sites may at first sight seem just incidental, but having an adaptable and well-behaved plant that stays neatly in place is probably the missing key to mitigating the harshness of many now barren lands. Vetiver hedges in this case would be deployed as vegetative shock troops to seize a botanical bridgehead on hostile lands and open the way for other species to follow.

It seems likely that the lines of solid plant cover will indeed help get the restoration process started. Already the effect can be seen in many places. In Louisiana, for example, barren washes quickly fill with native vegetation after vetiver hedges stabilize the area. In northern India, sodic wastes were turned into luxuriant forests once vetiver hedges were in place. And in southern India, forests have been seen to colonize hillsides after vetiver hedges provided some protection.

This particular vetiver use is hardly well known and is deserving of its own dedicated initiative. Research, testing and a comparison of experiences are all needed in a wealth of difficult sites. Globally important extreme soils to include are vertisols, laterites, saline and sodic types. The "laterite" that dominates the lowland tropics is an especially potent challenge. That particular soil, red in colour, very acid and high in soluble aluminum, a deadly toxin to most plants, has long been considered beyond the possibility of high-yield farming, but the fact that vetiver survives (even thrives) in laterite could turn out to be one of the great breakthroughs for tropical agriculture and forestry.

Combinations of vetiver hedges with appropriate leguminous cover crops that renovate infertile land between the hedges need especial consideration. That one-two punch, based on a natural succession of the vetiver pioneer and the nitrogen-fixing successor, should open the doors to routine development of many now unusable sites. The combination with laterite-tolerant leguminous trees, such as *Acacia mangium*, could also be a powerful intervention.

Taken all round the Extreme-Soil Initiative is a way to "sell" vetiver to a new set of clients for whom soil erosion is not a main concern. Examples are land-use planners, international donors, economists, policymakers, government administrators and others worried over population pressures and immediate food supplies. In principle, hundreds of millions of hectares of now unused lands could be rejuvenated to support more people and more crops. Turning wastelands into farmlands would, in addition, be a way to save more natural forests from slash-and-burn destruction.

The Water-Management Initiative

The fact that vetiver hedges are dense enough to dam up water is yet another distinctive feature. The effect is due to the plant's stems and myriad leaves, as well as to the soil and litter that collect behind a hedge. The effect is more sophisticated than people imagine; a vetiver hedge handles different depths of water in different ways. A modest, ground-level run-off hitting one of those hedges gets ponded, but a rushing torrent passes through with increasing ease as it rises past the point where the leaves splay outward. An established hedge seldom gets knocked down, and its variable-filter feature, damming up ground-level flows but progressively passing more water the deeper it gets, is an important one.

Professionals and policymakers involved in water issues are unaware that vetiver can help their efforts. This Water-Management Initiative needs to reach out and show them what they have to gain. Topics to highlight include the following.

Watershed management

By holding silt and water on hillslopes, vetiver hedges should be able to protect entire watersheds in the way that the original forests did. This would not only reduce soil loss and river sedimentation, but by keeping water on the land, vetiver would recharge groundwater supplies. Work in Malaysia shows that by using plants raised in pots, the hedges can become functioning barriers within weeks of being planted out. This holds the possibility of creating "instant" working watersheds over vast areas at modest cost. It would also mean that people might be able to stay living on the watersheds without severely affecting the area's vital hydrologic importance.

Waterway stabilization

Vetiver planted along streams, river banks, canals, drains and ditches can help keep out silt, maintain the flow and prevent the banks from being undermined. This means, among other things, that capital investments in water supplies will be protected and enhanced.

Reinforcing

This coarse grass with its roots like chicken mesh projecting several metres into the soil probably can strengthen earthen structures such as small dams and dikes. Following the disastrous Mississippi floods of 1993, it was reported that all levees protected by switchgrass remained unbroken. Vetiver should do at least as well because it is endowed with a better root and stem architecture for the task.

Sediment control

Waterside "walls" of vetiver hedge, grown on the banks of reservoirs, would provide ideal holding pens for dredge spoil. By allowing the water to filter back into the reservoir, these cheap, porous barriers would make it feasible to isolate the solids for economic handling by people or machines. Such self-rising silt-traps might help rescue reservoirs serving cities such as San Juan, Puerto Rico and Port-au-Prince, Haiti. Those reservoirs, along with many more in the tropics, are fast silting up and prematurely losing their capacity to hold water or generate electricity. Of course, vetiver hedges should also be put on the watersheds to stop the whole siltation process happening over again.

Engineering water flow

Vetiver hedges can be employed not only to retard run-off but to direct water toward, away from or through some given point. Hedges angled down slopes, for instance, would divert water away from sites such as unstable cliffs. For the cost of a few tillers and a planting effort, hydrologists and engineers could harness nature to achieve water shedding or water harvesting or other forms of water control.

Wastewater treatment

Probably there is no better species for stripping nutrients out of domestic and, perhaps, industrial wastewater. A native of a wetland environment, vetiver withstands long immersion. Hedges grown across or around man-made marshes would probably block the passage of solids, strip out dissolved nutrients and detoxify pathogenic microbes through aeration or detention. By providing simple, compact water-treatment facilities that require no chemicals or pumps, vetiver could create a new and cheap form of tertiary wastewater treatment for the countries of the "Vetiver Zone." In return, these wastewater treatment facilities could become vetiver nurseries. Fertilized by the wastewater nutrients, the plants should throw off tillers in abundance. Employing human wastes to grow vetiver for planting where it can do good for people and the environment is a new and especially elegant notion of recycling.

In summary, this Water-Management Initiative could elevate vetiver into a tool for providing more reliable water supplies, reinforcing earthen dams, protecting riverbanks, treating domestic wastewater and much more. In selling the idea to hydrologists, sanitary engineers, public health specialists and so forth, a few spectacular successes could make all the difference. The Panama Canal, for example, would be a great showcase. Today, ship wakes erode parts of the canal banks, but vetiver hedges would absorb the swells and allow ship traffic to speed up, thereby increasing the canal's throughput and economy. Moreover, contour hedges on the surrounding hills and mountains would retard rainfall run-off, recharge groundwater supplies and probably restore the Chagres River to reliable year-round flow as in the days when those watersheds were fully clad in forest.

The Pollution-Control Initiative

Although vetiver has many potential uses in pollution control, none is being vigorously developed or promoted. The initiative needed here is to reach out to governments, environmental scientists, industry and non-governmental organizations (NGO) concerned over cleaning up the messes that people or their institutions have left behind. Also, vetiver might be employed to prevent future messes from occurring or at least from spreading. Examples of what vetiver might help clean up are given below.

Underground flows

Surrounding polluted sites with vetiver hedges may well be a way to keep toxic compounds from moving outward underground. The massive, curtain-like "hangings" of interwoven roots seem ideally structured to filter out underground contaminants. If the plant can keep deadly pollutants corralled and unable to move outward and contaminate new ground, vetiver will have earned a place in everyone's gratitude.

Soil

Paul Truong's magnificent work in Australia has shown that vetiver is tolerant of high levels of arsenic, cadmium, chromium, copper and nickel. The plant therefore seems highly suitable for rehabilitating and maybe reclaiming lands contaminated by heavy metals, as well as perhaps by radionuclides and similar horrors resulting from mining, other industries, research facilities, landfills or illegal waste dumps.

Industrial spills

No one has reported trying vetiver hedges against spills of industrial liquids, but it seems to me that a series of these very dense hedges would provide a cheap and probably effective backup protection against small spills at least. It would hardly matter if the hedges died, they could be easily replaced. Even crude oil might be held back. Indeed the oil-soaked vegetation could be burned for furnace fuel.

Run-off

As mentioned above, vetiver hedges could block nutrient-laden run-off, which is a rising concern these days. Vetiver hedges could be especially useful as a "filtration barrier" around farms, industrial facilities, landfills and even golf courses. It could also be useful around ponds and marshes built to contain or detain run-off, notably stormwater from city streets.

Natural waters

Hydroponics might be a way to use vetiver hedges to filter dangerous materials out of surface waters. This is a speculative and untested idea but, as noted, the plant is at home in watery conditions. In one form of hydroponics the plants would be grown in an inert and highly pervious material through which the waters would pass. Coir dust, peatmoss, or perlite are possibilities. In another, vetiver might be grown with its massive roots dangling free in the water. This far-out idea, which works for other plants, requires something (old tyres perhaps) to keep the vetivers from sinking. Floating hedges might even be deployed across streams or canals to strip pollutants and dissolved nutrients out of the water flowing past. This water-borne process might even prove a convenient way of growing vetiver roots for oil extraction as no digging is needed, as the root ends are clipped when they grow too long).

Industrial wastewater

The possibility of treating human wastes in man-made vetiver-filled wetlands has already been mentioned. This non-chemical wastewater treatment also seems promising for cleaning waste products from aquaculture. It is already removing nutrients from trout-farm effluent in trials at a U.S. Department of Agriculture research facility in West Virginia.

Taken all round, this Pollution-Control Initiative opens up vetiver applications relating to some of the best funded areas of research, with billions of dollars being spent on pollution controls in the United States alone. But the use of the grass is not currently a part of the experts' thinking. To correct that, vetiver needs to be tested widely in polluted sites, and fast. A success or two could launch our grass into new, big-time and well funded applications. In fact it would transform the world's appreciation of the plant overnight. In people's minds, a tool for removing deadly toxic hazards is something quite different from a tool for controlling soil erosion on foreign farms. Such a change of attitude would help everything.

The Farmer-Support Initiative

Unless farmers deeply appreciate the plant and fully recognize that they are benefiting from it daily, we will always have to struggle to get vetiver hedges on the land. So, whilst emphasizing grand global problem-solving, such as mentioned above, it is necessary to get millions of farmers eagerly planting vetiver for themselves and selling the surplus. To assure this special efforts are needed to produce worldwide appreciation of the benefits to growers. Many farmers will not plant anything new just for erosion control, but they will eagerly tend a crop that provides income or makes their lives easier or more secure. Here are some features of vetiver that provide saleable products or a better life for farm families. *Handicrafts*: vetiver's bamboo-like stems are ideal for making baskets and other small saleable items. *Thatch*: the leaves make one of the longest lasting and most beautiful roofs. *Supplementary Feed*: although not a great feedstuff, vetiver is better than many give it credit. *Improved Crop Yields*: holding moisture back fosters better crop growth and helps keep wells filled. *Wildlife Controls*: pests such as rodents and Africa's grain-devouring quelea bird might be kept out of crops. The birds, for example, like to roost in blocks of tall grass, and there can be trapped in the darkness of night and used as food. *Mulch*: the leaves create a long-lived mulch that helps garden plants survive adversity. *Windbreaks*: standing up to 3 m tall, vetiver is ideally structured to resist the wind. *Boundary Markers*: several African nations recognize property lines demarked by vetiver because it stays in such a narrow band. It is a cheap way to stake out property rights. *Air-conditioning*: mats woven from vetiver roots are placed over window openings and doused with water to cool millions of India's houses. This both chills and perfumes the breezes passing through and the process may have much wider potential. *Ornamentals*: in Miami, vetiver plants are being taken up for their beauty and good behavior in the landscape. *Screening*: the tall, dense hedges are a way to provide a measure of privacy around houses, latrines, etc.

Moreover, circles of vetiver might be used to enclose compost piles, trash heaps, farm gardens, fish ponds and more. *Animal Protection*: corrals and shelters for small creatures such as chickens seem a possibility. *Traffic Control*: vetiver can be employed to direct where people and animals walk and where vehicles drive. For instance, it can keep them away from unstable banks or family gardens. *Weed Prevention*: the hedges are said to prevent creeping weeds, such as Bahia grass, from invading gardens. *Making Steep Slopes Usable*: hedges across slopes make it possible to work where now even standing is difficult and everything washes away with the rains.

All of these farmer advantages need to be developed and exploited throughout vetiver country. They should be brought together in extension literature. In this case, the publications might mention erosion-prevention, but their more immediate purpose is to stress benefits to the farmers' daily income and existence. In addition, commercial markets for vetiver tillers, handicrafts, thatch, "air-conditioning mats" and other products need to be advanced. Moreover, rather than establish centralized nurseries, a commerce in farmer-supplied planting materials should be encouraged.

The Disaster-Prevention Initiative

Given the deep roots, high tops and thick hedges, as well as the promise of practical large-scale application, it seems obvious that this grass could play a role in mitigating and, perhaps, preventing various natural disasters.

This topic, speaking technically, overlaps water management and soil-erosion control, but speaking in the political and humanitarian sense, the topic of disaster prevention makes vetiver of interest to different ministries and industries. Here the ultimate goal is not just to control water and retain soil, but to save lives and reduce property damage. The Disaster-Prevention Initiative, then, is a way to reach out to the worldwide insurance business, mortgage lenders, governments and more. Below are some possibilities where vetiver might make the difference.

Mudslides

The stiff, strong tops of vetiver hedges stop mud and debris from passing by. The massive underground walls of interlocking roots seem likely to stop slopes from slumping. The plant should operate on an essentially permanent basis to protect unstable sites from causing damage to property and people's lives.

Floods

It has already been mentioned that vetiver hedges may be planted in ways that rob floodwaters of the power to cause destruction. This, and the fact that the hedges can hold rainfall on the watersheds, should help reduce the devastation of flooding.

Fires

In South Africa it has been found that burning off the hedges at the end of the wet season results in a flush of growth that stays succulent through the subsequent dry months. The local insurance industry has accepted this band of green vegetation as an effective firebreak.

Droughts

By helping extend groundwater and surface water supplies in for instance, watersheds and reservoirs as mentioned above, vetiver should be able to benefit drought-prone areas.

Earthen structures

Some (many?) earthen structures are in danger of collapse. A decade or two ago a dam in the hills above Los Angeles burst, releasing a deluge that caused immense property damage and some deaths. Vetiver appears to have the potential to be an inexpensive reinforcement for such structures. Levees around New Orleans and along the lower Mississippi are likely candidates. Were they to break, the devastation could be catastrophic.

One cannot be certain about vetiver's utility in any of these undertakings, but the authorities charged with disaster prevention should be given a chance to put vetiver to the test. Whatever is done to prevent disasters such as floods will have to be done over vast areas, but vetiver seems more suited to large-scale application than other possibilities, such as those employing concrete and steel.

This use of vetiver in emergency management would come clear to the appropriate authorities and businesses if it were employed on some high-profile sites. An example might be Mt. Pinatubo in the Philippines where the massive landslides of volcanic debris are inundating towns and villages and forcing the relocation of whole peoples. In addition, disastrous mudslides have in the last few years caused deaths and/or destruction in Puerto Rico, Haiti, Leyte in the Philippines and Malibu in Southern California. All those locations seem ideal for vigorous vetiver growth.

In addition, the Mekong watershed might be tackled as an international vetiver-planting testbed. The idea would be to keep silt out of the river and future floodwaters out of people's houses. If the critical upland slopes can be returned to the hydrological state they enjoyed when fully forested then perhaps Thailand's terrible floods can become only a thing of memory. A similar, but even bigger, challenge would be the protection of Bangladesh from Himalayan floodwaters. Such a mission might seem too vast to be possible, but vetiver would be a better intervention to start with than anything else currently understood.

The Basic-Science Initiative

For all our experience, the truth is that vetiver specialists still do not know much about how the plant functions. Yet the workings of vetiver underpin everything claimed or envisaged. What makes this grass work so well at so many things? What makes it different from other plants? These and many more questions need urgent answering. (For references on vetiver anatomy, biochemistry and physiology see Chapter 2.)

In this Basic-Science Initiative the audience comprises specialists such as plant physiologists, microbiologists and agrostologists (grass scientists). The topics here relate to pure science, rather than strictly to practical affairs. Areas for investigation include *CO$_2$ Absorption*: in this era of the global warming scare, it is important to measure how much greenhouse gas vetiver stores in its massive roots. *^{13}C Absorption*: is vetiver, like corn, an accumulator of this uncommon isotope? *Taxonomy*: what exactly is the relationship between the sterile domesticated plants and the seed-producing wild ones? *Translocation of Oxygen*: rice survives in flooded paddies because it moves oxygen down into its roots. Vetiver also survives in paddies. Can it do the same? *Heavy Metals*: how well do pollutants move upwards from the roots to the leaves? Is vetiver a "super-bioaccumulator"? *Disease Prevention*: the plant is remarkably healthy, but we need to understand the basis of its resistance to such things as fungi and viruses. We also need breeding and selection programs because even a small and solitary disease outbreak could threaten the whole world's vetiver plantings. *Mechanism of Sterility*: why is the plant sterile? How reliable is that sterility? *Genetic Diversity*: what are the different types of vetiver? Are some better adapted for the various purposes than others? *Mycorrhiza*: these fungi that colonize roots probably are one of the keys to the plant's survival in extreme sites. We need to know more about such symbiosis (see Chapters 2 and 7). *Nitrogen Fixation*: does vetiver survive on barren sites because, like a few other grasses, it has a symbiosis with nitrogen-fixing bacteria? *Cold Sensitivity*: this is perhaps the biggest limitation for temperate-zone countries such as the United States. Can it be reduced or overcome? *General Tolerances*: what are the theoretical limits to drought, waterlogging, and toxic conditions? What can be expected in practice? *Mechanism of Hedge Formation*: why do the plants in a hedge tend to interlock when most grasses stay in separate clumps? *Dwarfing*: can shorter plants be obtained? *Root Growth*: just how strong are those reinforcings in the soil?

With topics such as these it is necessary to reach out to scientists in the appropriate fields and show them how by applying their expertise to vetiver they can produce data of global importance. This is one area where vetiver specialists have the possibility of finding research partners likely to devote time and energy without much cajoling. This is because in the grass family vetiver falls between sugarcane, sorghum and corn, which means that it probably has much to contribute to the better understanding of those billion-dollar natural resources. Researchers studying the basics of sugarcane, sorghum and corn, are therefore natural allies of vetiver specialists.

Conclusions

Breaking up the subject into these seven initiatives, can help generate funding, collaboration, innovation and new progress. More importantly, perhaps, it will inject vetiver into different disciplines, bringing in people with a range of special skills and backgrounds and insights. No longer will vetiver be the exclusive intellectual property of agriculturists; sharing our excitement will be environmentalists, chemists, engineers, hydrologists, and more. By this process of reaching out, vetiver champions can speak in seven voices, in seven forums, and stimulate outward momentum in seven directions. Also, the process will give a feedback from seven different outlying visions that are now glimpsed only vaguely, if at all. That will help better exploration of this immensely useful plant, and most of all that will help the people of the world.

The big buzzword today is "sustainable." Everything, it seems, must to be sustainable. Millions of dollars are being spent in arguing what is and what is not sustainable, but the debaters are not even considering the solid substance that lies at the heart of sustainability: soil. If one cannot hold onto soil one cannot hold onto anything. Without stable soil, "sustainability" is just an empty word that will disappear from everyone's vocabulary. But sustainability is actually a very valuable concept and an important challenge. Vetiver is the key to unlocking the heart of sustainability. Vetiver can literally make sustainability sustainable. Vetiver, as you have seen throuhout this book, can hold soil.

According to the Centro Internacional de Agricultura Tropical (CIAT) the annual global increase in atmospheric CO_2 is about 20 billion tons. By considering that producing new vetiver plants at the rate of 100 million a year provides 500 million kg or 500,000 tons of "atmospheric cooling" benefit the annual output of a single station growing vetiver plants will absorb the equivalent to one forty-thousandth of the world's CO_2 surplus. This ability of vetiver to fix greenhouse gases and turn them into underground solids is a further measure of the huge potential of this fascinating species.